ACCELERATING INNOVATION

IMPROVING THE PROCESS OF PRODUCT DEVELOPMENT

ACCELERATING INNOVATION

IMPROVING THE PROCESS OF PRODUCT DEVELOPMENT

MARVIN L. PATTERSON
WITH
SAM LIGHTMAN

Library of Congress Catalog Card Number 92-34006
ISBN 0-422-01378-7

Printed in the United States of America

Van Nostrand Reinhold
115 Fifth Avenue
New York, New York 10003

Chapman and Hall
2–6 Boundary Row
London, SE1 8HN, England

Thomas Nelson Australia
102 Dodds Street
South Melbourne 3205
Victoria, Australia

Nelson Canada
1120 Birchmount Road
Scarborough, Ontario M1K 5G4, Canada

16 15 14 13 12 11 10 9 8 7 6 5 4 3 2 1

Library of Congress Cataloging-in-Publication Data

Patterson, Marvin L.
Accelerating innovation / Marvin L. Patterson.
p. cm.
Includes index.
ISBN 0-442-01378-7
1. Production engineering. 2. New Products.
3. Production planning. I. Title.
TS176.P367 1992
658.5'75—dc20 93-34006
CIP

This book is dedicated to Hewlett-Packard Company engineers throughout the world. May your lives have balance—as professionals, as spouses, as parents, as citizens, and as individuals. May you always have fun at work.

Contents

Acknowledgments

This book is the accumulated result of years of work and countless discussions with virtually thousands of people, both from Hewlett-Packard Company and other companies around the world. It is impossible to recognize here the many people who have helped shape these thoughts. I would like to recognize a few, though, who have been particularly valuable to me in this effort.

I would first like to recognize the contributions made by Charlie Elman during our early work on R&D productivity concepts and training. Charlie gets specific credit for Figure 3-1. He and I received an assignment some years ago to develop productivity metrics for the R&D function and were shocked to learn that we couldn't even define R&D productivity let alone measure it. This work started then. John Fenoglio soon joined us and has helped a great deal in moving it along. I especially have appreciated John's ongoing friendship, moral support, and counsel as this work has progressed. I also would like to thank Norm Johnson for keeping us in touch with reality along the way. Norm joined the team soon after John and has added the pragmatic balance that we needed.

Professor William Ruch of Arizona State University and Professor William Werther, Jr., of the University of Miami were instrumental in getting our efforts on the right track. They have served us as consultants and mentors over the past few years, and their ideas are inextricably woven throughout this work.

A couple of my friends, Dave Ricci and Dick Levitt, have provided stimulating conversation, excellent pointers to essential reading material and specific ideas throughout this effort. Figure 3-4 reflects Dave's ideas on measuring processes. Dick invented the project-progress-rate metric that has spread widely throughout HP and appears in this work under another name in Figure 3-6.

I am particularly indebted to the management staff of HP's Medical Products Group for sharing their ideas about the product development process with me. Their challenging definition of the innovation cycle is given in Figure 1-1 and has led to essential insights that are reflected throughout this book. Opportunities to exchange ideas with people like this have provided the intellectual fuel that has moved this project forward.

The entire Corporate Engineering team deserves recognition for their creativity, enthusiasm and hard work toward making difficult dreams come true. Their experiences in the trenches with HP R&D project teams have provided essential lessons that are reflected in this work.

About a year ago, a consultant, Philip R. Thomas of Thomas Group Inc., prodded me to write my thoughts down in a book. When I pled lack of time, he brushed my objections aside and said, "Get a writer. That's what I do." What a novel idea! If he had not given that push when I needed it this book may never have happened. I am grateful to him for that and for the expertise that he has shared in his own books and lectures.

John Young, HP's president and chief executive officer, also deserves recognition. In the fall of 1990, John asked me to deliver a talk on reducing time to market to executives from another company who were due to visit in a few days. This assignment triggered discovery of the information assembly line metaphor and the subsequent evolutionary train of thought that has provided the basis for much of this book. More recently, John reviewed this manuscript and provided valuable suggestions that have improved the quality of this effort measurably.

I would particularly like to acknowledge my own project team, Sam Lightman and Alethia Green, who got this book to the publisher in only five months—only one working day behind our original schedule. This book simply would not have happened without them. Alethia created the graphics after-hours and on weekends. Sam transformed the ideas into readable form and, in doing so, saved the readers from my own sleep-inducing style. There are a lot of Sam's own thoughts in this book as well. He has been a constant cheerleader and friend throughout this process, and his professionalism has made this project both easy and enjoyable.

Finally, I would like to thank my wife, Mary Lou, who has provided both encouragement to me and loving concern for the words in this book. Her companionship has given me the solid base at home needed to support a project like this. Her professional attention to the words presented here has helped keep them straight and true.

Barney Oliver, retired vice president of R&D at HP, said once that people work at HP because it's a chance to accomplish great things with their friends. Great work goes on all over Hewlett-Packard Company in moving the new product development process forward, and it is truly a pleasure to be involved in it with my friends.

Marv Patterson

Preface

No existing market share is safe today, no product life indefinite. Not only in computers and clothing, but in everything from insurance policies to medical care to travel packages, competition tears away niches and whole chunks of established business with the weapon of innovation. Companies shrivel and die unless they can create an endless stream of new products.

ALVIN TOFFLER, *POWERSHIFT*

Everywhere one goes in the United States these days, business people seem to be busier, working harder, and struggling to stay competitive. Marketplaces are becoming increasingly global, offering new opportunities and challenges. Broad worldwide markets bring the promise of greater revenues but also changing ground rules and fierce competition. In this dynamic environment, some U.S. businesses are doing well, others are just holding their own, while many are losing the battle and going under.

As the urgency to develop new products increases in this turmoil, all available engineers must be thrown into the fray. Although they work intensely and for long hours, the actual project work often is rushed, inefficient, and uncoordinated. As a result, engineers go home at night frustrated, wishing they could take more pride in their work. Businesses correctly perceive the need for faster new-product development cycles, but in their haste to solve short term business needs, they fail to make the fundamental long-term investments that will improve their capacity to respond more quickly. So, in spite of more intense efforts, competitive positions often stagnate or even slip.

As Toffler suggests, new product innovation is the key weapon in this global battle for business success. The good news is that innovation is a cultural strength in America. We have an unparalleled history of successful innovation that has touched nearly every aspect of the lives of people worldwide. The bad news is that our innovations have yielded less economic return in recent years, and our standard of living is declining as a result. The responses by competitors to our innovations often have been quicker and more effective than our own. If our products simply alert competitors to a new opportunity, they reap the lion's share of the rewards by following our initial products with more effective offerings.

How can we convert our inherent innovativeness more effectively into competitive strength? How can we get products right the first time? How can we stay competitive without burning out the engineering work force? Is there a way of looking at product innovation that allows us to succeed by working smarter instead of harder?

We will find answers by exploring the benefits of faster product-innovation cycles. These benefits are so significant that accelerating innovation is essential to any successful long-term business strategy. This strategy requires management of continuous improvement in new-product generation processes, those cross-functional activities that systematically add value to information.

We can add value to product-development information as we transform well-proven best practices found in manufacturing operations around the world. In addition, we can support the re-engineering of product-generation work with principles that affect not only specific day-to-day actions at the project level but also long-term strategic planning for the entire business enterprise.

While this book examines engineering processes, the language is suitable for non-engineering readers. (Several equations in Chapter Six may be ignored without undue loss of impact.) The intended audience is the product generation management team in any business, and the principles outlined apply to everyone involved, ranging from the manager of an individual project to the company CEO, as well as to individual engineering contributors.

Any business that needs to develop new products can apply the principles immediately, even though the book does not offer quick fixes. These companies can ensure competitive new-product activities only through strategic investment in long-term improvement of the cross-functional, product generation process. Furthermore, the commitment to this investment must be permanent and never-ending. To decide that new product activities are good enough and that further improvement is unnecessary is to set the clock ticking on the eventual demise of the business. Inevitably some competitor who does commit to continuous process improvement will overtake and obliterate any business that stops improving.

Although improvement programs for product development come with no guarantees, successful outcomes are highly probable if programs are based upon sound principles that are applied well. More stories of such successes would be a valuable addition to the book, but few are available. Even though the concepts described are valid and well-proven in other fields, they are not yet applied widely in the product development process. The author hopes that this work will focus attention on the importance of the product generation process and stimulate the widespread adoption of process improvement principles.

Marv Patterson
Los Altos, California

CHAPTER 1

The Importance of Innovation Cycle Time

But the true "hollowing" of America is the loss of technological and innovative leadership, supposedly America's long-term competitive advantage, because of a stubborn refusal to face up to the core of the problem—long new product development and introduction cycles.

STALK AND HOUT, *COMPETING AGAINST TIME*

Everything you've heard is true. There is indeed a war on out there—a global war in the marketplace, with far-reaching economic consequences for companies and countries alike. For manufacturing enterprises, the air might just as well be full of bullets as innovations, given the near-lethal effect of each new product announcement. The issue here definitely is survival.

While the war metaphor may be troubling, it is not inapt. The war jeopardizes standard of living, jobs, livelihoods, and self-esteem both individually and nationally. Many U.S. businesses are not holding their own relative to the many other economic forces of the world, and if they are to survive, let alone prosper, then their armies are going to have to do better.

The armies of economic warfare comprise the sales people on the front lines who compete for customers, along with the divisions to the rear that support them—customer service people, order processing people, manufacturing people, marketing people, and so forth. Most removed from the heat of the battle are the battalions of product development people who create the weapons of war for this global economic conflict. Products that satisfy customer needs are the weapons that the product development people deliver. How quickly they develop the weapons, how competitive those weapons are, how supportable—the quality of the weapon and the timeliness of its deployment determine the outcome of the battle.

Many companies are inclined to function in the Same Old Way (SOW), the way things have always worked in the past. One by one, the economic armies of the United States, for example, fighting with inferior economic weapons, have lost their territory. Steel, autos, consumer electronics—the long and ignominious list of market battles lost has been covered exhaustively in the business press and needs no further elaboration here. The conclusion, however, is clear: Doing business the Same Old Way is a guaranteed losing strategy, because if the SOW continues long enough, some competitor will certainly come up with a better way and put those who are unwilling to change out of business.

Surprisingly, the product development community is particularly inclined to do things the Same Old Way, often for a variety of reasons. Sometimes the people in the product development lab are so pressed to get new products out as fast as possible that they feel they cannot take time to investigate alternative processes. Every available engineer must be thrown into the competitive fray. In other areas, the development team is so buffered from the front lines that it lacks a clear perspective on both the importance of its role in each battle and the need for urgency and more streamlined development processes. The players do not even know there is a war on until their project is canceled or their company folds. By then there is no time to react. One would expect product development teams, the traditional source of new technologies and new product ideas, also to demonstrate innovation in development processes. This is, however, rarely the case.

The time to react is long before the livelihood of the product development function is ever threatened. The battle strategy for product development has to be: *to bring new technology to bear on customer needs faster and more effectively than the competition.* Both characteristics are equally essential—faster and better. And if a company is trying to come from behind, as is too often the case, then it has to improve faster than its toughest competitors if it expects ever to overtake them and win.

The Japanese have demonstrated that it is possible to improve faster than competitors. Starting from so far behind after World War II that many people thought their products were ludicrous, they instituted a rate of productivity improvement on the order of 20 percent a year, compared with 1 percent to 3 percent in the United States. They absorbed W. Edwards Deming's teachings on improving assembly line efficiency into their culture and made process improvement a way of life. Now, to catch up in any given industry—or stay ahead in the few in which the United States still has superiority—it's crucial to improve faster than they do.

INNOVATION CYCLE TIME

Over the past decade, U.S. enterprise has concentrated on the area most susceptible to improvement, the manufacturing arena. American companies slowly and painfully have learned the lessons of total quality control, computer-integrated manufacturing, and just-in-time inventory control. Where they have not made progress, however, is in the area of product development. Yet, product development is one of the key pieces to the whole puzzle of competitiveness.

There are two basic objectives of any business: to satisfy customers and to make a return on investment. In manufacturing, ROI (return on investment) means creating cash flow back for the money invested in a new product. The ratio of return to the initial investment is the scorecard; higher is better.

The one factor that has the most impact on both objectives, customer satisfaction, and ROI, is the time it takes to develop and introduce a new product, measured from the time the opportunity for it occurs. To visualize the nature of that impact, consider Figure 1-1, Innovation Life Cycle.

The time T_0 is the time the opportunity for a new product occurs. It is a philosophical point in time, not usually discernible: the moment

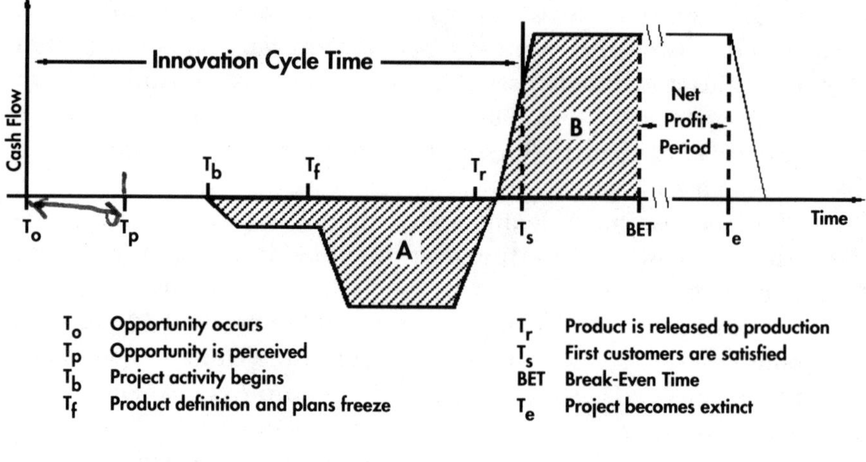

FIGURE 1-1.

Innovation Life Cycle.

when an emerging technology overlays a customer need and triggers a new product possibility. This definition is, of course, variable with circumstance. A new T-shirt product, for example, becomes viable when the home team wins the championship. Clearly the technology to make T-shirts with any conceivable message existed before this particular event. Customer needs change, however, as a result of the local victory, and an opportunity appears. Opportunities occur this way as well in the conventional world of manufacturing. Often though, opportunities occur when a new technology makes possible new solutions to existing customer problems.

Product innovation cycle time is the time between the moment when the window opens and the moment the first customers are satisfied. The opportunity occurs and generally is followed by some delay until time T_p when it is perceived. It is business' job to reduce that delay time to a minimum and get a product into that window as quickly as possible.

Figure 1-1 implies that the opportunity always precedes development, but this is not always true. Occasionally product development can actually start ahead of the opportunity. In the computer printer business, for example, the opportunity for color printers was limited until color copiers became pervasive in the office environment. Even with multiple copy output, color documents were restricted to local distribution. An original sent elsewhere could not be duplicated conveniently, and if information was carried in the color, the document was rendered useless by black and white duplication. With the spread of color copiers, color printers, for which product development had already taken place, became viable. Of course, dramatic reductions in product costs have helped as well.

Thus, if engineers have an idea that anticipates the occurrence of a market opportunity, they can begin development in advance. With any luck, by the time the opportunity occurs, they can launch the product that exploits it.

By definition, the moment the opportunity occurs, people would buy a product to match it if one were available. Every month's delay in the introduction of a suitable product is a month of lost revenue. In addition, if a product arrives in the marketplace after one or more competitors already have launched theirs, not only has time been lost in the market window but market share as well. To be preceded to market by a competitive product puts a company in an unenviable, defensive position. So business success is a steadily decreasing function of product introduction time: the later a company introduces a product, the lower the possibilities of business success.

There is an interesting exception to this case, in which one competitor arrives on the scene behind another, but with a clearly superior product. Many Asian consumer electronics manufacturers have built successful businesses by following pioneers with this "me-too-but-better" strategy.

By and large, however, most companies do not have the luxury of that kind of approach to their markets. The Japanese have demonstrated a superb capability for taking the ideas of others and expressing them in a world-class, manufacturable, high-quality embodiment. Americans, on the other hand, have demonstrated the capacity for real innovation. It is up to American companies to deliver these innovations to their customers in packages designed and manufactured so well and priced so competitively that no room is left for imitators.

Returning to Figure 1-1, the market window is not known *a priori*. The moment of opportunity may or may not be discernible, but the extinction time (T_e) is always obscure. Extinction time depends to a large extent on imponderables: how technology changes over time, the competition's response, a company's own next move. The only controls a company has over extinction time are the choices it makes early in the definition of the product: the target market, the feature set, the operational technology, the manufacturing technology, the price point, the distribution channels, the customer support system, and so on. The tougher it is to compete with a product, the further the extinction time extends.

Once the plan freezes, the extinction time freezes as well, although at an unknown future point. Assuming that it is the best implementation of the contribution that can be developed with the current processes and technology, there is nothing left on the table for other companies. In order to complete, they must invent a fundamentally better contribution with new technology or processes.

In some cases, however, extinction time is determined by the product's introduction rather than its definition. Real-world experience suggests there is often something into which the competition can sink its teeth. The reaction of the competition will thus eventually make the product obsolete. Alternatively, when project teams finish the new product, they are free to go off and invent its successor. In either case, the extinction time would appear to be determined more by the product introduction time than the product definition.

A good example occurred when engineers purchased a unique, very expensive device to perform non-intrusive measurements on moving machinery. The device performed well enough, but its construction was appallingly bad. It was unreliable and required frequent service and adjustment. Out of curiosity, the engineers disassembled it and found that the design, mechanical layout, and fabrication were very poorly executed.

The product in this case created a tremendous opportunity for the competition. A competitor could have duplicated its contribution using existing best practices and very quickly put a more reliable adaptation on the market at a significantly lower price. It was clearly a product whose introduction could have triggered its extinction. Patents might have provided some protection from this likelihood, but these all too often prove to be a flimsy defense against competitors. The best defense is a product implementation that is simply tough to beat.

In any case, the product life cycle is a finite window determined by a multitude of influences. For the purposes of this discussion, it is assumed that product extinction is determined when the product definition freezes.

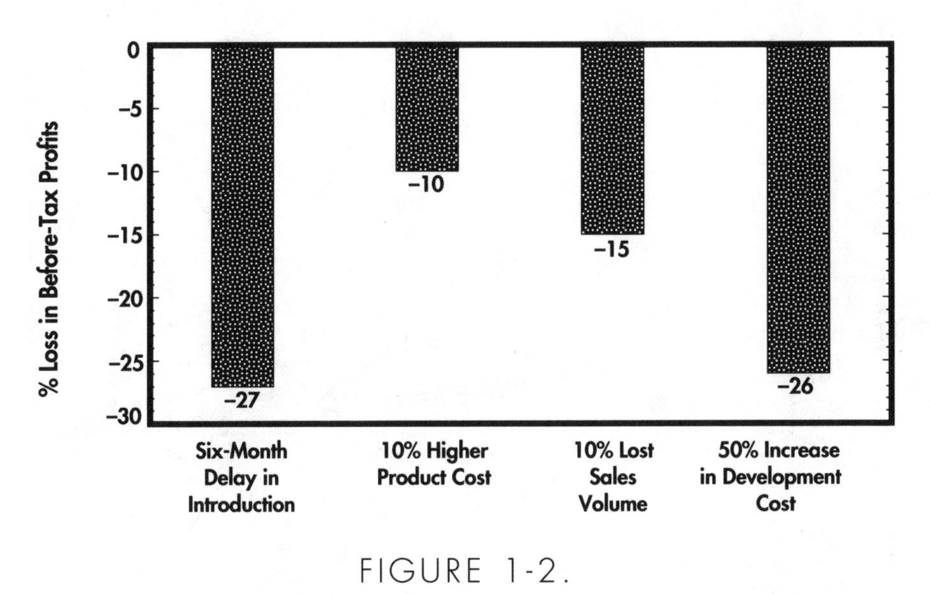

FIGURE 1-2.

Sensitivity of Profits Over Product Life.

THE COST OF TIME

The object of product development then is to identify opportunities, define the most competitive product possible and get the product to market expeditiously. Once the product definition is frozen, the primary factor that determines return on investment is the speed with which that product comes to life. Every month the product launch slips takes a month away from the product's return window. Looking at the diagram in Figure 1-1, product release time (T_r) moves to the right, but extinction time (T_e) does not move, and return shrinks accordingly. (In reality, it's worse than Figure 1-1 indicates, because the diagram makes no allowance for market share, the loss of which further reduces the amplitude of cash flow.) There is no way to recover that loss. The product has a finite period of opportunity, and if it misses a month in that period, revenue is essentially lost.

The impact of product development cycle time on profitability has been addressed by Smith and Reinertsen.[1] Four scenarios were analyzed for their impact on before-tax profits:

1. A six month delay in product introduction
2. A 10 percent increase in product cost

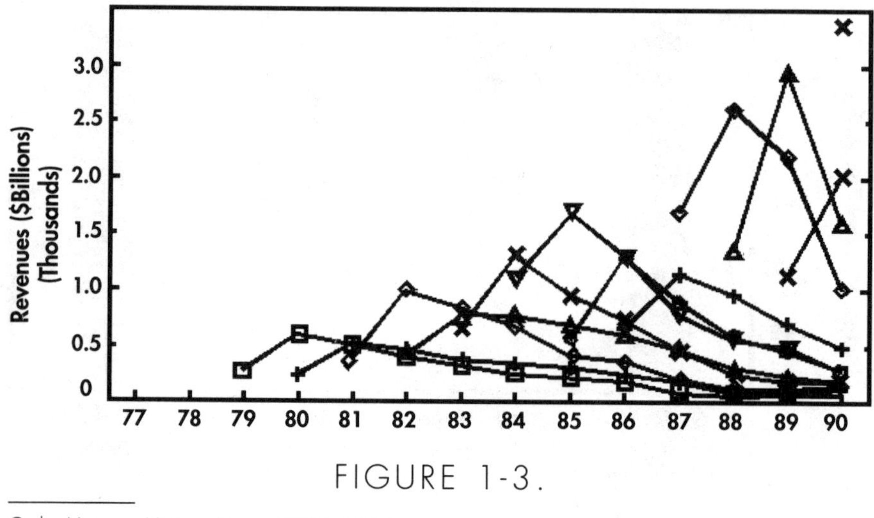

FIGURE 1-3.

Order History—Vintage Years 1985–1990

3. A 10 percent decrease in unit sales due to reduced product performance
4. A 50 percent overrun in development cost.

The sales window assumed in the analysis was about five years. The results are illustrated in Figure 1-2. Delaying the product has a greater impact than the other three scenarios. In fact, this data implies that increasing the development investment by 50 percent to keep the project on schedule would be a good business decision. Businesses rarely take such radical steps to keep projects on schedule, but because of the incredibly high value of time in the market, perhaps they should. In this example, a lost month in the market is worth about $470,000.

As implied by Figure 1-1, the impact of project delays on investment return increases as market windows get shorter. A look at HP sales history makes the five year window assumed by Smith and Reinertsen seem somewhat enviable. Figure 1-3 shows aggregate sales of all HP products by year of introduction, the so-called "vintage year." The curve that begins in 1979, for instance, gives the total sales over time for all products introduced in that year. This somewhat complicated collection of curves illustrates a couple of points. First, for each vintage year, sales start out low, peak in the second year, and then trail off over time; there is a consistency in the sales waveform. Second,

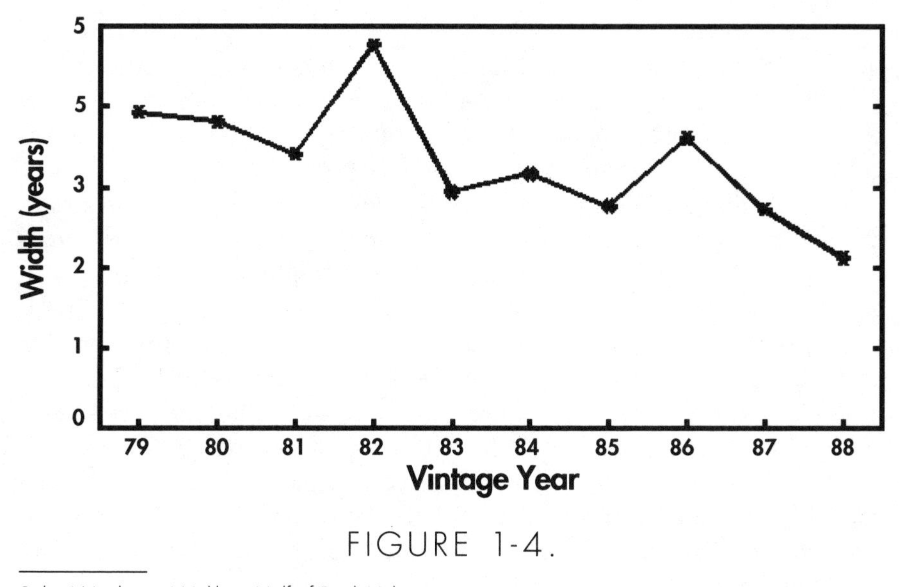

FIGURE 1-4.

Sales Window—Width at Half of Peak Value.

there is a rather dramatic increase over time in the revenue created by new product introductions. Finally, the width of the sales envelope for each vintage year seems to be getting shorter with time. To explore this effect, the width of each vintage year sales envelope is plotted at its 50 percent point in Figure 1-4. While this data is not directly equivalent to market window, it is related. This graph implies that, since 1979, the average time in the market for an HP product has decreased by a factor of two. The urgency of getting products to market quicker is increasing, and the cost of time is growing rapidly.

While Figure 1-1 is obviously a simplified model, it sufficiently reflects reality to communicate some powerful concepts. The most critical period to business success is depicted on the diagram as the period between the time the opportunity occurs and the time the first customers are satisfied. While that period is not always measurable, it is the time most profitably minimized. Success derives from defining the best possible product and moving it to market quickly.

Most conventional definitions of development cycle time start the clock when the project begins. The cycle time defined here begins at the time the opportunity occurs (T_o) instead of the time project activity begins (T_b) to highlight the fact that some of the most important opportunities for business lie in the dead time between opportunity and project definition. For every month's delay there is exactly one month of lost sales. That month is much easier to recover than a month lost in product development time. Shortening the time it takes to perceive opportunity has the same business impact and is significantly easier to accomplish.

Reducing that dead time is best achieved through effective strategic planning, market research, and technology research. Every business should—must, if it expects to be successful—maintain an ongoing program which scans technological innovations, tracks competitors' activities, and actively pursues the perception of customer needs and expectations. This program should be operating in the background at all times and should be constantly comparing market needs with emerging technologies to perceive opportunities that are relevant to the business. This ongoing program is a relatively low-cost effort, probably the most cost-effective approach available for reducing the elapsed time between the time the opportunity occurs (T_o) and the time the first customers are satisfied (T_s).

The purpose of this book is to look at the processes of product generation and how they relate to the waveform of Figure 1-1 and then to determine how managers can evoke behavior from their organizations that optimizes this waveform. One mandatory behavior is to establish the necessary background processes such as market research and technology scanning and then foster their success and continuation in spite of exhortations to "downsize" from elsewhere within the organization. The confluence of available technology with market readiness triggers a business opportunity. If a company misses enough of them, it endangers the continued existence of the business itself.

MORE ADVANTAGES OF SHORT CYCLE TIMES

Higher Sampling Rate

Occasionally it is possible to apply engineering principles to the product generation process with advantageous results. One such principle is the Nyquist rate as applied to market opportunities.

Systems which cannot be tracked continuously must be monitored through sampling techniques. Nyquist developed a mathematical relationship which demonstrates that to recover the information in any given waveform, the sampling rate has to be at least twice the highest frequency in the waveform.[2] The higher the rate of sampling, the greater the likelihood of recovering all the information. But the minimum rate is 2 times. Thus if a company has a system that changes monthly, it must sample at least twice a month to be confident that it is on top of developments within that system.

Applying this to marketplaces, it becomes clear that to stay abreast of market opportunities, the higher the sampling rate, the more successful a company will be. If the business is fast-moving and dynamic—such as personal computer products, for example, where product lifetimes are as short as 12 months—market sampling needs to occur at a much higher rate than for products in the lumber industry. Having a tighter handle on the marketplace helps significantly to shorten the time between the moment the opportunity occurs (T_o) and the time the opportunity is perceived (T_p).

One of the most powerful sampling techniques available is new product introduction. Every time a company introduces a new product into the marketplace, it receives a rush of feedback that is unavailable to competitors. Customers describe what they like about the new product and what they do not like. They apply the product in ways never dreamed possible, and these new applications suggest new opportunities for future products. The more often a company introduces new products, the better it learns the dynamics of its marketplace.

Accordingly, shorter product introduction times provide a higher sampling rate that gives a better capacity to track the marketplace. Assuming that a company is engaged with its market and providing incremental improvements in its product line on an on-going basis, if it introduces products at 18-month intervals while competitors introduce at 24 months, it will learn more about the marketplace faster than they. Higher frequency product introductions give an advantage in the marketplace.

Better Product Definition Stability

An additional advantage of shorter product introduction cycles is definition stability. Projects often slip because the definition changes in mid-stream. A product that takes 48 months to develop is almost certain to go through definition changes in the process. On the other hand, moving products out the door at the rate of one every 18 months gives each a fair prospect of becoming the product it was originally intended to be. The likelihood of definition change increases dramatically with product development time; shorten the development time and the prospects for definition stability improve. The waste associated with a mid-stream change of direction is reduced, and return on investment improves commensurately. The amplitude of the positive side of the waveform increases and the magnitude of the investment side decreases.

Leadership Reputation

Shorter cycle times also provide an intangible benefit in the form of perceived market leadership. New products gain the attention of the marketplace and improve the company's image as a market leader. They make it appear more responsive to customers' needs and hence the one they look to first when they are considering a purchase. An interesting example of the value of this phenomenon occurred when HP surveyed its customers on the subject of facsimile machines. In a blind questionnaire, HP asked its printer customers whose plain-paper fax machine they would buy if they were contemplating a purchase. The list included the names of the major players in that market, and even though HP did not have a fax machine at the time, HP's name was on the list as well. The majority of respondents chose HP. HP has since introduced a line of plain paper fax machines that leverage its inkjet technology and are having a significant impact on the market.

Increased Organizational Learning

According to de Geus[3], "The ability to learn faster than your competitors may be the only sustainable competitive advantage." *Because product development is a process and is repetitive,* there is ample opportunity to learn and improve. But much of what might be learned is lost if product cycle times drag on indefinitely. Over a 48-month product development cycle, the problems encountered at the beginning that caused all the slippage have been forgotten by the time of the post-mortem analysis. If the product cycle is 18 months, however, there is a good chance of recovering all the information associated with it for the purposes of improving on the next cycle. Conclusion: the shorter the cycle time, the higher the learning rate. Since, as has been said at the outset, a company has to learn faster and improve faster than its competitors if it is to regain or retain its markets, short product cycles are an important factor in the process.

The value of shorter product development cycles is thus established as consequential to a whole range of worthwhile organizational objectives, from increasing return on investment to improving competitiveness. The remainder of this book will focus on the approaches necessary to accomplish ever-decreasing innovation cycle times.

NOTES

[1] Smith, Preston G. and Donald G. Reinertsen. 1991. *Developing Products in Half the Time.* New York: Van Nostrand Reinhold.

[2] Oppenheim, Alan V. and Ronald W. Schafer. 1975. *Digital Signal Processing.* New York: Prentice Hall.

[3] de Geus, Arie P., Head of Planning, Royal Dutch/Shell. March/April 1988. *Harvard Business Review.*

CHAPTER 2

Process Improvement Considerations

Implementing such far-reaching innovations involves wrenching changes at all levels of the organization and requires an extraordinary commitment by corporate leaders.

DERTOUZOS, LESTER, AND SOLOW, *MADE IN AMERICA*

The goal of a process improvement program must be to create and maintain a rate of improvement greater than that of the competition. To improve the performance of an organization's product development community, the first step is to recognize that product development is indeed a *process*. The prevalent notion that product development is somehow all black art and invention is a considerable impediment to change. Translating a market opportunity into a new product requires perhaps 15 percent invention. The remaining 85 percent of the work involves previously learned processes that often are undocumented and undisciplined.

Accept the premise that most product development work is process rather than invention, and it's possible to predict the outcome of each activity as the result of the associated process. For example, if engineers are struggling to find the information they need to exercise their knowledge, waste is taking place. The process is broken because the information is not there when the engineers need it.

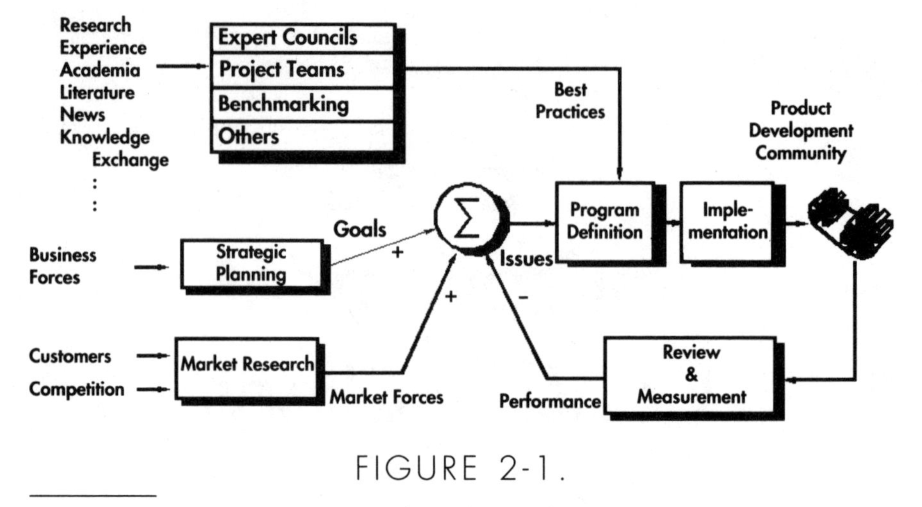

FIGURE 2-1.

A Continuous Process Improvement System for Product Development.

Figure 2-1 addresses in a theoretical context the elements that must be in place within the organization to accomplish the goal of more competitive new product development. The product development community comprises not only the research and development people, but marketing people, quality assurance people, and manufacturing people as well. The intent of Figure 2-1 is to illustrate how to positively change the behavior of this cross-functional team in order to achieve more competitive performance.

Managers will recognize Figure 2-1 as a schematic representation of the Total Quality Management process. Engineers will see Figure 2-1 as a feedback control system, in this case acting on the performance of the product development community. It does indeed obey some of the same laws of control systems theory to which an electromechanical control loop would respond, and in fact the similarities run deeper than one might expect.

The drivers of this system are the forces entering on the lower left: the competitors' capabilities, customer needs and how they are changing, and the company's business needs. These variables are compared with current performance of the product development activity to determine competitive issues.

Issues are identified by analyzing the differences between these market forces and the product development community's performance as determined by the review and measurement process. For example, if

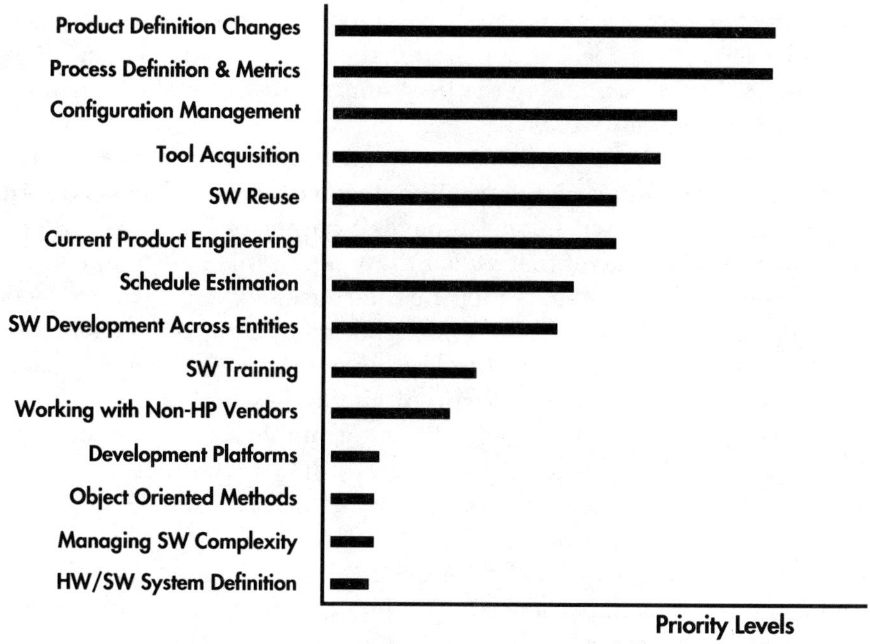

FIGURE 2-2.

Prioritized Issues.

the competition can introduce a new product in 18 months while it takes another company 24 months, subtracting one performance level from the other provides a six-month difference, which then becomes a competitive issue.

Once the issues become clear, their relative importance must be determined: which ones are the most important to business success and which most impede competitiveness. Figure 2-2 provides a hypothetical list of prioritized issues that might appear on that line. Note that the issue set will be dynamic, changing with time. As an organization, the competition, the overall marketplace, and the business environment all shift in response to the latest inputs, the issue set should respond accordingly.

Before a company can begin to define programs to deal with the issues, however, it needs to bring one more element, called "best practice," to bear on the issues. Best practice means the most effective business techniques known that can transform a particular set of issues into competitive advantage. These techniques might include organizational design, processes, and methods or tools that relate to a particular problem area. With best practice and the issues clearly stated, it is possible to begin assembling the programs needed to make an organization more competitive.

Where does best practice come from? It derives from the process of filtering the flood of information that constantly flows into an organization, as shown in the upper left corner of Figure 2-1. Some very small percentage of that huge inflow represents best practice.

"Benchmarking" represents one filtering mechanism that extracts best practice. In their study of American manufacturing capabilities, *Made in America; Regaining the Productive Edge*, Dertouzos, Lester, and Solow state, "A characteristic of all the best-practice American firms we observed, large or small, is an emphasis on competitive benchmarking: comparing the performance of their products and work processes with those of the world leaders in order to achieve improvement and to measure progress."[1]

By surveying a variety of companies, in businesses both similar and different, new approaches to product development problems can be found. For example, understanding how a consumer-product company measures the reaction of customers to a new toothpaste might help an electronic-instrument manufacturer solve its product definition problem. Furthermore, extracting the common elements from a variety of successful practices can lead to a new process that is even better.

Benchmarking is not, however, a universally admired concept. Philip R. Thomas claims that companies that depend upon Benchmarking to determine best practice are doomed to mediocrity.[2] This, he feels, is a "me-too" strategy that leads to goals for the future that are only marginally better than what others are currently able to achieve.

Instead, he emphasizes a concept he calls "entitlement." Each process has an inherent performance limit set by the physics of the process. In manufacturing, for example, this is the bare minimum machine time required to make a part. Other non-value-adding factors, such as handling, transport, change-over, and so forth, are not included. Thomas urges goals based on the entitlement computed for each process. He claims that the performance of companies that set goals in this way is often five times better than that achieved through Benchmarking.

Other mechanisms for extracting best practice from incoming information include project teams and expert councils. At Hewlett-

Packard Company, for example, a software technology council meets quarterly to discuss software development problems and practices. Information from outside sources is systematically reviewed for applicability. The collective expertise on this council represents an essential competitive asset for the company. Project teams at HP routinely conduct an analysis of completed projects to identify opportunities for improvement. This information is used to advance processes used in subsequent projects. Sharing of best practices that result occurs through a project management council, which also meets quarterly, and through an annual companywide project management conference.

Dertouzos, et al, made some interesting observations about best practice:

> *We call attention here to six key similarities among the best-practice firms: (1) a focus on simultaneous improvement in cost, quality, and delivery; (2) closer links to customers; (3) closer relationships with suppliers; (4) the effective use of technology for strategic advantage; (5) less hierarchical and less compartmentalized organizations for greater flexibility; and (6) human-resource policies that promote continuous learning, teamwork, participation, and flexibility.*
>
> *...our best-practice firms put particular emphasis on* simultaneous *improvements in quality, cost, and speed.*

With best practice and the issue set as inputs, improvement programs can now be defined. These should be viewed as internal products and services to be delivered to an internal customer base. Best product definition methods should be used, delivery channels considered, and cost of both delivery and ownership identified. Marketing plans should be made that create both awareness and demand when the program is ready for introduction. Internal products and services are often the hardest of all to sell and so require the best marketing practices available.

Once the programs have been defined, they must be implemented in order to change the behavior of the product development community in specific and directed ways. We will elaborate on implementation below. To complete the discussion on Figure 2-1, note simply that a change in the behavior of the product development community will very shortly begin to show up in the review and measurement area as improved performance. For example, reducing the disparity between the innovation cycle times of the competition and the company will result in a modification of the issue set. As more and more of the original issues are resolved, new issues will emerge and move up the priority list. They will require new best practice to assist in defining new programs. The whole sequence goes around in a closed loop that continuously improves the processes affecting the performance of the product development community.

Attempting to change the performance of the product development community without tracking its behavior with sufficient frequency—or measuring the wrong variables, drawing the wrong conclusions, and consequently implementing the wrong programs—can disrupt its performance seriously. The review and measurement feedback element of the loop is therefore critical to its successful operation. Different kinds of review are required to track different aspects of the product development community's performance in an adequate fashion.

There is extensive literature on the subject of reviewing and measuring behavior in the corporate environment, albeit not often particularized to the product development area. One of the favorite review approaches in HP is management by wandering around, MBWA[3]. It serves well as a review process because managers who walk around not only see what their immediate subordinates are doing but what the people who report to them are doing as well. MBWA is one indication of a healthy product development environment.

There are also more formal kinds of reviews—project milestones, annual reviews and so forth. They are all attempts to meet an important criterion for change: understand the reality of the business and be able to describe it accurately.

Ultimately, the review and measurement mechanisms must be designed to track the necessary variables at the intervals appropriate to the frequency of change. A congruence is required between the measurement systems used versus the response time of the community and its ability to change behavior. Projects which change on a weekly or monthly basis naturally will require more frequent review than the community's product strategy, which may not change more than once or twice a year.

Similarly, sampling competitor performance, customer needs, and business needs all must be done in a way that makes the informa-

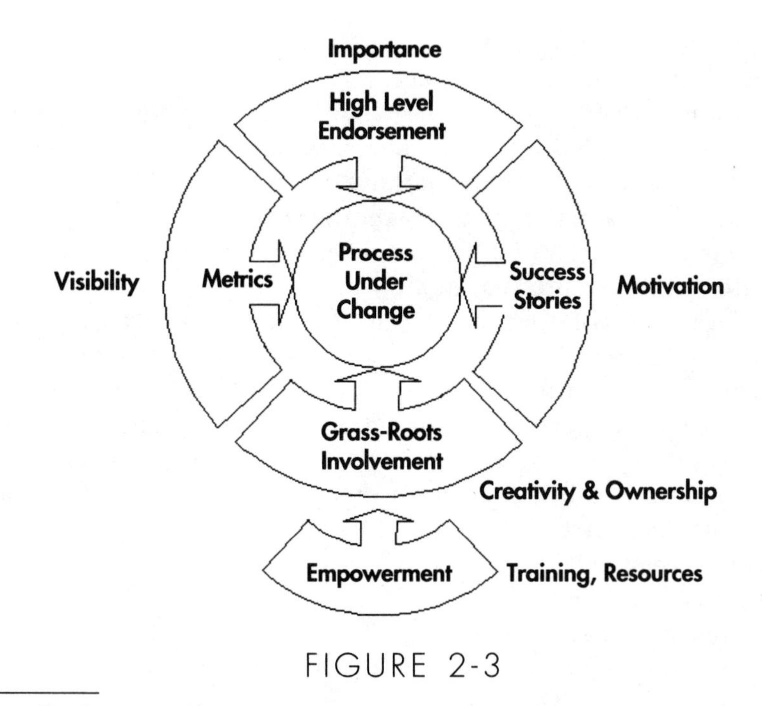

FIGURE 2-3

Factors That Encourage Change.

tion available for comparison in the summing junction real and timely. The insight provided by the Nyquist criterion regarding sampling rates (noted in Chapter 1) allows a company to begin to design its business systems appropriately. Market research probably would intuitively design the correct frequency for a customer-needs sampling program, but it is comforting to know that someone has put a theoretical boundary on the task.

IMPLEMENTATION

It is one thing to define the perfect program, quite another to implement it in a way that fosters the desired behavior change. Figure 2-3 provides some insights into how this might be accomplished. It represents a distillation of information garnered from consultants' presentations, publications, and the author's experience in the area of productivity management. The net message of Figure 2-3 is that all these factors need to be in place in some reasonable balance to get the desired behavioral changes to occur.

Although Figure 2-3 is accessible from any point, it's convenient to begin with high-level endorsement. High-level endorsement creates an environment that makes the desired change important if not mandatory. A good example of the effect of high-level endorsement occurred in 1981 when CEO John Young decided that HP could transform product quality into a competitive differentiator. He launched what was called the 10X hardware quality goal: a ten-times reduction in the annualized failure rate per $1000 of list price, to be accomplished within ten years. By the time of the general managers' meeting, about a year after he had announced the original goal, he had obtained enough information to know that not much progress was being made in the desired direction. His solution: a scorecard. While he was addressing the meeting, a list went up on the screen behind him with the names of all of the manufacturing divisions in HP, beginning with the one who had made the most progress and ending with those that had moved a little or not at all in the desired direction. He then made a point of recognizing openly the efforts of those at the top of the list. His approach had the desired effect. Essentially, with one stroke he had invoked all the factors required for a desired change in behavior. He displayed the success stories implicitly; he involved the grass-roots (in this case, the managers); he provided a metric by which all could determine their progress, and he certainly demonstrated high-level endorsement of the desired change. The goal, in fact, was achieved, and just about on schedule.

As stated by John Young, the 10X goal was sufficiently demanding to require across-the-board changes in the operation of every division. Inspecting for quality was out; designing for quality was in. Manufacturing for quality was in. Few activities within the company escaped scrutiny or process dissection as a result of the new objective.

Any time a process change is mandated, a concomitant change in the behavior of the grass-roots people involved is required. At the CEO level, the grass-roots people are the division managers who will champion the new goal. From the division down, however, employees at every level must be involved in the design of the program to change the process. According to classic TQM theory, the grass-roots must describe their existing processes, analyze them, determine the obstacles to improvement and then become the owners in defining the changes

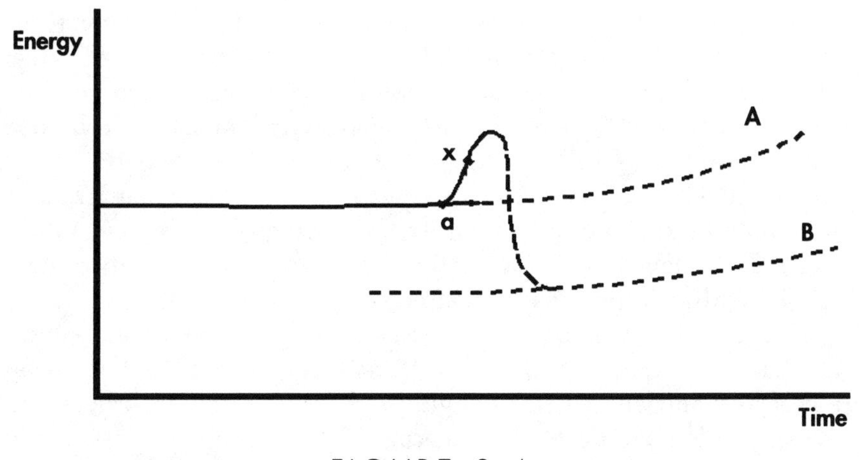

FIGURE 2-4.

Energy Barrier to Change.

required to meet the new goals. Ownership of the change elicits creativity. Since the people engaged in the process already know more about that process than anyone else, they also will know best what it takes to fix it. Grass-roots involvement, then, is critical to the success of any program aimed at changing the behavior of the people performing the process.

What happens when the grass-roots have no input into program design? Consider the typing pool experience. At one company (not HP), managers decided to try to improve the productivity of the typing pool. After insufficient consideration, and no consultation with the participants, they settled on keystrokes per unit time as the metric by which to measure productivity and its improvement. They invested in a keystroke counter on every keyboard. Still no consultation with the typists. Next, someone from outside the pool came around daily to record the numbers on the counters, thereby lending importance to the metric. The result: one day the counter reader came by at lunch time to take readings and found all the secretaries eating at their desks—chatting, reading or talking on the telephone—each with a thumb pressed firmly on the space bar.

Clearly, setting metrics and establishing their importance without grass-roots endorsement will generate a change in behavior, but not necessarily the change desired.

Empowerment fosters involvement. Training and resources help, but the people involved must also have the ability to do something different, make their own rules, and experiment without risk to career and credibility. Without empowerment, involvement is uncertain and change is unlikely. HP established "productivity managers," provided them with training and resources but failed in some parts of the company to empower them to make changes to their assigned process. Where they lacked empowerment from their high-level superiors, the experiment failed. Elsewhere in the company, where empowerment existed, significant progress was achieved.

Empowerment should be done in a prototypical manner, to minimize risk to the company while the effects of the changes to the process are digested and analyzed. Changing a deeply ingrained process involves considerable risk to all concerned, for all the obvious reasons. If the change fails to produce the desired results, the company has lost productivity and gained nothing but experience. ("Experience is what you get when you didn't get what you wanted"—Anonymous.) The people involved in the experiment are apt to become gun-shy, particularly if they are punished for taking the risk and failing. (This point will be considered later.)

Not only has the failed experiment cost productivity, but it has incurred a tremendous energy expenditure on the part of the individuals concerned—for naught. It is not uncommon, however, that when the team stays the course and expends yet more effort, the desired results are achieved. This phenomenon is displayed in Figure 2-4, which illustrates the energy barrier to change.

Curve A describes doing business the Same Old Way. Some added energy costs are incurred over time just to continue in the SOW, because of baggage accumulated over time, primarily in the form of support required by products already in the marketplace.

Curve B is a suggested new paradigm with the potential for significant competitive advantage, whether through productivity gains, quality improvement or some other development that enhances the company's position in the marketplace. If achieved, it will cost the company significantly less energy over time. Success stories and existence proofs may or may not exist to support the claim, but the promise is there. However, any success stories that may have emerged will not match the environment in question exactly, so there is still a sizeable risk associated with the process changes necessary to achieve the goal.

The path from Curve A to Curve B is hazardous and tortuous. It requires an investment in the learning curve. It also requires the intestinal fortitude to slow down existing projects while the new process is primed. Once over the hump, the organization will drop down very quickly to Curve B and begin realizing the gains it sought at the outset. But the risk is that the investment will not pay off and all the effort will be wasted.

The manager at Point a has a decision to make: invest a little and continue in the SOW, which has benefits over the short term, or invest a lot in the pot of gold at the end of a very long rainbow.

Within every organization of any size, there is some 5 percent of the population that is ready to take the risk. The trick is to locate that pioneering spirit, the sponsor environment that really wants to go for it. This group is eager for improvement; they are tired of being beaten up by the competition, or they are determined to stay on top of their market and recognize the value of the new paradigm in their efforts.

It is incumbent upon the champion to partner with the pioneers and help them over the energy barrier. When they've moved to Point X on the curve and suddenly realize that they've expended all the energy it took them to get there—and they still have not achieved their goal—it is time for the champion to roll up his or her sleeves and help them hurdle that energy barrier to minimize the impact in any way possible. Otherwise, other organizational factors will prevail, and they likely will drop back to the SOW curve. That drop is the point at which the potential for change evaporates. This energy barrier is repeatedly the cause of failure as people try to leap it and fall back, or choose not to try it at all.

The whole performance is analogous to chopping wood with a dull ax because it takes too much time and energy to sharpen the tool. Woodsmen will expend incrementally greater effort to achieve the same production. If they stop to sharpen axes, productivity suffers a brief interruption, but they very quickly regain the lost production and then increase productivity significantly. To overcome the reluctance to stop chopping, they have to be convinced that the sharpening process will indeed improve their axes, and thus their productivity.

Throughout the entire experience of overcoming the energy barrier to change, wisdom dictates that the pioneers record as much as possible about productivity (or competitive position or whatever is pertinent to the operation) both before and after the process change.

What were the costs? What were the gains? What was the ROI? The pioneers then can be used as success stories to promote the desired process change elsewhere in the organization where, presumably, similar efforts will bring about similar gains.

NOTES

[1] Dertouzos, Michael L.; Richard K. Lester, and Robert M. Solow and the MIT Commission on Industrial Productivity. 1989. *Made in America, Regaining the Productive Edge.* Cambridge, Massachusetts: The MIT Press.

[2] Thomas, Philip R. 1990. *Competitiveness Through Total Cycle Time.* New York: McGraw-Hill Publishing Co.

[3] Peters, Thomas J. and Robert H. Waterman. 1982. *In Search of Excellence.* New York: Harper & Row.

CHAPTER 3

Designing Metrics

What kind of behavior
are you trying to encourage?

CONVERSATION WITH
PROFESSORS WILLIAM RUCH AND WILLIAM WERTHER.

Metrics are a double-edged sword, representing almost as much potential hazard as benefit to the user. They can affect significant changes in behavior, but the nature of those changes also can be beyond the wildest dreams of the metrics' initiator.

Some years ago, engineers at HP attempted to design a system of productivity metrics that could be used to track improvements in the R&D process. It was intuitively clear that they could not discern progress without metrics, so they took on the task of creating them.

They quickly discovered that they could hardly describe R&D productivity, much less define it. Once they overcame that problem, they then developed a set of measurements they thought were appropriate and subsequently presented them to a worldwide meeting of R&D managers. The responses were universally pejorative.

Why? Apparently, had HP implemented this particular set of measures, the character who would have done best with them was the kind of manager most disliked: the person who takes an ultra-conservative approach, overestimates all projects, and does a good job of meeting schedules but a poor job of creating new products per unit investment.

The engineers suddenly saw metrics with new eyes—not as a means of measuring performance but as an impetus to encouraging behavior. What people measure, how they measure it, and how much importance they place on the metric will determine what gets done, how it is accomplished, and, just as important, what gets ignored.

Remember the example in the last chapter of secretaries with the keystroke counters? Behavior is clearly a function of the property being measured. The metric had better be conceived very carefully.

The secretaries' response to the keystroke metric falls under the heading of "malicious compliance." Another example closer to home is the case of the time-to-market metric applied by one high-tech company. The managers of an electronics manufacturer decided that the key metric to success for their business was to establish clear time-to-market goals for the various product activities and to attach recognition and reward to performance on the time-to-market metric. This would seem to make perfect business sense. What could be better than pushing time to market? And the rewards were substantial. Stock options and salary increases were affected directly by the measured results achieved on time-to-market. The problem that occurred was that time-to-market is only one of a number of metrics that are needed to ensure business success. And in fact some clever people in charge of new product introductions came to realize that time-to-market was really being measured against new product numbers introduced and the time between introductions. They hit upon the practice of relabeling existing products and publishing data sheets with new product numbers but often not really changing the designs that had been sold earlier. By manipulating the product number and data-sheet publications, they were able to make apparent progress in time-to-market look as good as they chose. This practice went undiscovered for several years while everybody involved reaped great rewards in stock options and salary increases. The practice was stopped only after it was discovered that no real contribution to the customer was being made.

In another example, one of HP's operations launched a concerted effort to reduce time-to-market by increased emphasis on the metric plus actually increasing R&D expenses as a percent of sales. Indeed, development cycle time decreased but sales did not increase. All that happened was that products turned over a lot faster. There is an important difference between simply driving down time to market and encouraging development behavior that results in real business success.

THE KEY ATTRIBUTES OF METRICS

Given the potential for unforeseen consequences, it is clear that appropriate metric design is critical to the success of any improvement effort. What are the key attributes of an effective set of metrics? There are four:

Relevance—The metric must provide clear information focused on factors important to the task at hand. Some programs impose metrics that require the gathering of exhaustive data that are only partially relevant to the business at hand. A well-supported and effective set of metrics should in each case be clearly relevant to an important aspect of the function being measured.

Completeness—The set of metrics together makes visible all important factors with balanced emphasis. Leave one out, and the system will trash that omitted parameter in ways that improve the others. For example, make time-to-market the most important measure, while forgetting to include a metric for product quality and watch what happens to quality. Every important aspect of the operation under scrutiny should be measured. The goal is a minimal number of metrics, but nothing should be omitted that is critical to success.

Timeliness—Timeliness is a function of how quickly the business can change. Metrics on any business activity must be sufficiently real-time to enable decisions that relate to the actual current state of the business at hand. Measures need not render the current state precisely, but they should be timely enough that decisions based on them are effective and correct. The Nyquist criterion discussed in Chapter 1 applies here as well. If a measurement can change weekly, that function should be tracked about twice a week; if it changes monthly, then check it twice a month. Based on the nature of the particular function being measured, the timeliness of the measurements taken should be adjusted as necessary to assure that the quality of decisions based on those measures will be high.

Elegance—The concept of elegance should not be overlooked. Metrics can be a serious burden to an organization, creating significant overhead. Metrics require effort to gather, report, compile, track, and archive. Whenever measurements are implemented they will impose a burden on the work force. The purpose of a set of metrics is to acquire insight into the actual performance of a business process. Insight is the

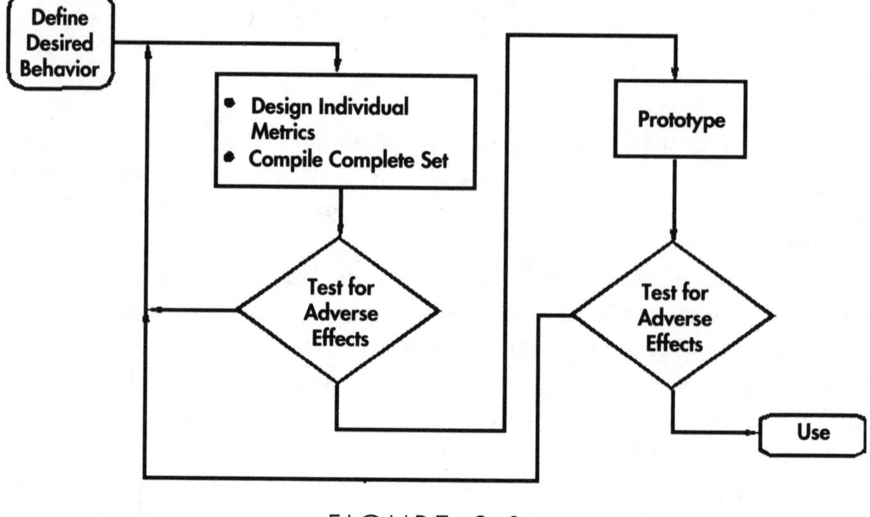

FIGURE 3-1.

Metrics Design Paradigm.

goal, not data or elaborate reports. If the metric set is designed with elegance, it will achieve a maximum level of insight with a minimum amount of data.

THE METRIC DESIGN PARADIGM

There are some key elements in the design paradigm for metrics (Figure 3-1) which must be considered. Keeping both the desired attributes and the behavioral consequences of metrics clearly in mind, the first task in designing metrics is to define the behavior that needs to be encouraged—a not insignificant challenge.

In both cases described in the anecdotes above, the metrics were in a sense pulled out of thin air. They seem fine, but in reality they began with a very limited view of the desired behaviors that management sought to achieve. And in fact, because of the lack of completeness in both cases, the behavior that was actually fostered was certainly not what was desired. So to design a set of metrics requires, first and foremost, a clear definition of the desired behavior. Some approaches to this requirement are considered below, but for the moment suffice it to say that this task turns out to be surprisingly difficult.

With desired behavior completely described, individual metrics must be designed with the key attributes in mind. Each metric should be tested at least mentally for relevance, completeness, timeliness, and elegance, and then the method by which the metric is actually gathered and applied should be adjusted appropriately. This mental test is applied to the entire set, and the set is adjusted accordingly until it passes the full attribute test.

Having established an apparently workable set of metrics, it is wise to move cautiously with them, preferably into a prototypical situation. Pick some reasonably controlled environment, actually implement the measures, and then watch for adverse results, including the diabolical "malicious compliance" effect. Since the proof of their appropriateness is in actual use, some measures will doubtless need to be fine-tuned. Go back through the loop as often as necessary to obtain the desired behavioral change. And finally, when the set satisfies expectations, apply the metrics to the rest of the organization where appropriate.

Distinctions must be made among *process metrics, engineering performance metrics,* and *management performance metrics.* Where processes are concerned, once again the criterion is desired behavior. The primary interest is in measures that indicate the state of the process and the results that it produces.

Processes have various attributes called "state variables," analogous to the state variables of a physical system. A spring-mass system, for example, can be characterized by the states of each of the masses of the system. These states are the positions and the velocities of the masses in the system. It's possible to predict how that system will behave if we know the parameters of the system and the state variables at any given time. Processes such as product development have an analogous set of state variables. If they can be established, they will serve as *process metrics*, to be tracked for the purpose of understanding and controlling the process.

Engineering performance metrics concern such properties as expertise, quality and quantity of work, judgment, teamwork, and "robustness" of the engineer. Engineers with robustness have the ability to stay on track and do what makes sense regardless of the influences that tend to distract them, managerial or otherwise. Again, these measures reflect desirable behaviors found in the best engineers.

Managerial performance must be measured in still other ways. In particular, the quality of decisions that managers make is a key measure of their performance, a subject considered more extensively below. Other measures of managerial performance include their leadership skills, their process expertise and how well they apply it, their ability to manage change, their business judgment, and another intangible called "asset stewardship." Managers are handed some critical assets of the organization to manage. They thus become stewards of those assets. They can choose to exploit them in favor of achieving other goals, or they can try to develop those assets and make them even more competitive. So asset stewardship is a key to managerial performance. Again, each of these qualities relates to desired and expected managerial behavior.

Evaluating managers' performance by the results they achieve is a natural tendency. Results are usually measured by the performance numbers: financial targets, schedules, market share, growth, profitability, and so on. However, these measures reflect management quality only to the degree that the manager has control over current results. In many situations, current results are only loosely related to recent actions of the manager in charge. Other factors such as market forces, competition, and decisions made by previous management can have a much greater impact on today's situation. In product development, the recent actions of a manager will have an impact months, or even years later. Undue emphasis on current numbers can lead to promoting the guilty and punishing the innocent.

Measuring a manager's results and using them to judge the quality of that manager is logically suspect. The error is compounded by the long time constant inherent within many organizations. For example, imagine a situation in which a manager is appointed to supervise an organization that has been well managed in the past. He takes control of an excellent work force, inherits a potent product strategy, and begins with a competitive edge in the marketplace. Unfortunately, the quality of management deteriorates abruptly under the new manager who makes poor decisions. The working environment deteriorates, and the best people leave the organization. Nevertheless, the results achieved by the organization do not change immediately and in fact continue on several years before the decline becomes obvious in financial terms. In the interim, the manager assigned to this exemplary organization looks great because he is being measured by results. He gets promoted out of this job and into the next. The manager who inherits the organization from him is unable to fend off the inevitable decline and eventually takes the punishment for the loss of momentum, and in fact the loss of competitive stature. Result measurements actually seem to be extremely poor indicators of management quality in this case.

An organization that insists on attaching such measures to management performance can expect a management culture to evolve in which risk is avoided at all costs. An independent consulting firm, the Strategic Decisions Group, of Menlo Park, California, puts forward the contrasting point of view that managers should be rewarded for the quality of their decisions rather than the outcome achieved by those decisions. This method of reward will ensure that appropriate risk-taking behavior can survive.

Consider another hypothetical case, in which a manager is given a set of choices. On the one hand she can invest $60 with 100 percent probability of getting a $70 return. Alternatively, she can invest $60 with the prospect of getting a $500 return 50 percent of the time, and $0 the other 50 percent. Which decision should the manager make? The best decision obviously is to gamble on making $500 half the time. That has an expected value of $250, based on a probability of 0.5 x $500, the outcome if the experiment is successful. So half the time, a manager who makes that choice will make a return of $500, with an average return of $250. For a $60 investment that is a much better return than a $70 payback 100 percent of the time.

If an organization judges managers' performance based on outcome as opposed to decision quality, however, the first time that managers who take the gamble come up empty-handed, their futures at that organization are in jeopardy. They simply will not survive in the no-risk culture. And over time, the organization that rewards strictly on results will drive the risk-takers right out of its ranks. On the other hand, those that always pick the sure thing, the $70 return for the $60 investment, will look like good solid managers. And those people will stay with that organization forever, whereas those who stick with the sure thing are exactly the folks the organization wants its competitors to hire. For entrepreneurial spirit to survive in the organization, measure decision quality, and put less emphasis on results. They inevitably will follow consistently good decisions.

Measuring decision quality, however, is more difficult than measuring results. Most financial systems are tuned to reporting results. Measuring results often becomes a simple matter of reading the bottom line and comparing it to expectations. Measuring decision quality means reviewing decisions of lower level managers and applying judgment. Is this the best that could have been done under the circumstances? Did this person use available information and a reasonable process to make the highest quality decision possible? This judgment implies a more personal contact with what is going on at lower levels. Higher level managers with good judgment and intuition have a real edge here.

Product development metrics are needed at several different levels. Project managers need information about the flow of activities within the product development process—such as the rate of progress on software module design, or the turn-around time and yield of printed circuit board design activity. Functional managers of product development need information about the states of their projects. Are projects on schedule? Do they have the required resources? Have they encountered unexpected problems? Finally, general managers need information about the portfolio of new product activities and about how well the cross-functional product generation process in the business unit is working.

While no attempt will be made here to provide a complete set of metrics for each of these levels, examples are given below, both of the descriptions of desired behavior and some useful performance measures.

DESIRED BEHAVIOR FOR THE CROSS-FUNCTIONAL PRODUCT GENERATION PROCESS

A candidate list of desired behaviors for product generation activity within the business unit is as follows:

1. Competitive development process with a high rate of improvement
2. Aggressive improvement in Break-Even Time (BET)
3. Effective support of other organizational elements
4. Effective post-release support of products
5. High productivity in delivering products to market

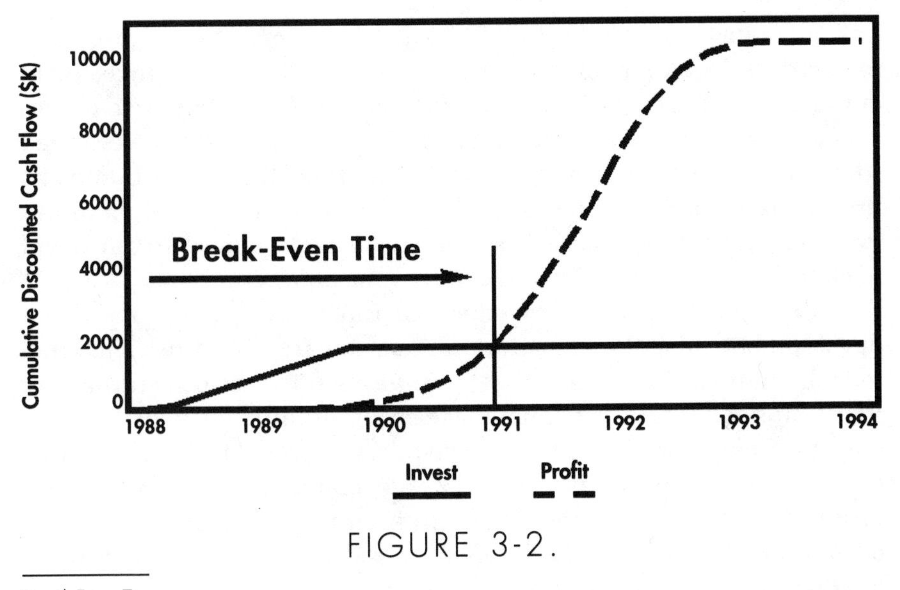

FIGURE 3-2.

Break-Even Time.

6. Competitive long-range product and technology strategies
7. Effective balance of resources between long-range opportunities and shorter-range product activities
8. Aggressive improvement in the competitive stature of the engineering work force.

To elaborate on the above list:

1. *A competitive development process with a high rate of improvement.* As discussed earlier, the development activity is perhaps 85 percent process and 15 percent invention. The part that is process should be viewed and managed as process, that is, subject to continuous process improvement strategies. To stay competitive, a business must maintain a continuously high rate of improvement in the development process. Accordingly, business unit managers must understand that they own the process, and the desired behavior from them is that they continuously improve this process. (See desired behavior for the product development process, below)

2. *Aggressive improvement in Break-Even Time.* BET is an interesting metric. Essentially it measures the time from the beginning of product development work until the product has been introduced and

has generated enough profit to pay back the investment originally made in its development. Figure 3-2 compares the original investment curve against the eventual profit-return curve. When those two curves cross, Break-Even has been achieved. BET is customarily measured from the time of the product launch. At Hewlett-Packard Company, product launch is often designated the "I-to-L" or Investigation to Lab Prototype review, the point when managers make a hard commitment to staff the project and develop a particular product.

BET emphasizes improvement of three key factors in product development, all at the same time, all in a constructive direction. First, for the company to break even, the investment must be returned, so BET helps focus the development team on holding costs down. Second, it stresses profitability. This requires a product that meets real customer needs and has a competitive manufacturing cost. Those profits depend on sales, though, so an improved BET also requires an efficient introduction into an effective sales channel. And third, it emphasizes doing all of these activities fast, more quickly than the competition. Achieving all of the desirable results simultaneously leads to shorter BETs and a more competitive product generation activity.

The three factors that are emphasized in trying to improve BET are the investment level, the profitability of the product, and the time it takes to move the product to market and get a return flow started from the marketplace. The emphasis on time, investment, and profitability are all pushing product development teams and the product development process in the direction of increased competitiveness.

Unfortunately, it turns out that BET does not satisfy some of the key criteria that were listed above. In particular, timeliness is not really sufficient in the BET metric. The ultimate value for BET does not register until a significant amount of time has passed after the product development process has been completed and the product has been released to the marketplace. So analyzing BET performance will not encourage meaningful decisions about a process while it is unfolding.

Further, BET should not be used as a single figure of merit to apply across a portfolio of products. If a company averages its BETs and one product fails to break even, the averages are skewed, and no meaningful information results from an aggregated BET metric. A scatter-gram of BETs for a portfolio, though, is useful. It is a very good metric for describing the desired behavior a company is trying to foster within its product development process. Furthermore, it is widely used within HP to assess the viability of individual projects before they are fully staffed.

3. *Effective support of other organizational elements.* Product development does not occur in a vacuum; it occurs within the larger organization. Other parts of the organization depend upon the performance of the product development teams. Manufacturing, for example, depends a great deal on product development to design and introduce products that can move smoothly into the manufacturing plant and be compatible with most if not all of the existing processes. Marketing and customer support both depend on product development for information needed to prepare support for new products required by the marketplace. So another element of desired behavior for the product development function is sensitivity to and support for the needs of other parts of the organization.

4. *Effective post-release support of products.* It is useful to view the product development team as functioning like a subcontractor working within the organization. This subcontractor makes commitments and then attempts to carry out those commitments, such as respecting the budget, delivering agreed-upon functionality, and meeting delivery schedules.

Among the functions or deliverables that compose these commitments are, for example, a product's manufacturability, its reliability, and perhaps its localizability into foreign marketplaces. These deliverables should be clearly articulated as goals early in the product development activity. Once product development is completed and the product is in the marketplace, the product development team should be, as with any subcontractor, responsible for meeting the commitments made. If the team members failed to meet those commitments—if the reliability, for example, is not what it was targeted to be—then the product development team should continue to support that product even after it has been released to the marketplace, until the design of the product has been modified adequately and the implementation is sufficient to meet all the original objectives. In a sense, post-release support is nothing more than a typical subcontractor's guarantee—in this case, provided by the product development team—that they will indeed fulfill agreed-upon commitments.

5. *High productivity in delivering products to market.* The product development team members are stewards of a significant amount of cash, human resources, and plant and material assets of the organization. They have the responsibility to use those re-

sources to deliver products. Naturally, they should do that job with as much efficiency as possible, without waste and without depletion of those resources.

6. *Competitive long-range product and technology strategies.* One thing an emphasis on BET might do is push people to focus on short-range projects that have a quick return on investment. Indeed, to shorten BET, a company simply can choose projects that can be done quickly and realize a fast cash return. This focus tends to pull effort away from the longer-range products that might move a company into a new area or that might be absolutely essential to the long-range future of the organization.

A balance between short-term and long-term goals should be reflected in the long-term product and technology strategies of the organization. The product development function is responsible for providing a critical influence on both product and technology strategies. It supplies leadership that helps to draw the organization into the right channels for future success.

7. *Effective balance of resources between long-range opportunities and shorter-range product activities.* Again, long-term must be balanced with the short-term. The product development function is responsible for the management of a portfolio of product activities. Management of that portfolio includes deciding what should be in it and how resources should be allocated across that collection of project activities. The effective balance between long- and short-term is a function of the larger organization, of the particular business, so there is no single right answer here. But clearly, the product development function should have the responsibility of making effective investment decisions between long-range, high-payback opportunities and shorter-range projects, perhaps along the lines of product-line enhancement activities.

8. *Aggressive improvement in the competitive stature of the engineering work force.* The engineering work force should be viewed as one of the most competitive business assets. The competitive stature of this work force is a vital element in the success of the business. A desired behavior of the product development function, then, is to actively develop this asset and not to deplete it.

There is a real temptation when other factors are measured and emphasized, such as time-to-market, to trade off the health and well-being of the engineering work force in favor of better performance against other metrics that are more visible. Often the engineering work force and its competitive stature are never monitored, never tracked, and hence are not visible. Overworking the engineers and eliminating developmental activities such as education can become routine. Abusing the engineering work force to the point where its competitive stature is significantly damaged is, in fact, quite possible in some organizations. A complete set of metrics therefore must not neglect the health and well-being of this asset.

DESIRED BEHAVIOR FOR THE PRODUCT DEVELOPMENT PROCESS

The development process checklist contains six items:

1. The product under development is consistent with organizational priorities, product strategies, and quality objectives.
2. Project commitments are consistently met, exceptions characterized, and contingency plans executed quickly.
3. Other operational departments are kept informed and brought on line smoothly.
4. The design maximizes use of standard parts and processes.
5. Product implementation is world-class and discourages competition.
6. The expected return on investment is high.

Going back through the list:

1. *Product development is consistent with organizational priorities, product strategies, and quality objectives.* Product development teams not only understand the priorities, strategies, and objectives of the organization before they launch a product, but they include commitments in the product definition to achieve the quality objectives of the organization and to provide a product that makes a substantial contribution to the overall strategy of the organization.

2. *Project commitments are consistently met. Exceptions are characterized quickly and contingency plans are executed.* In the product development process, project teams make commitments and then keep those commitments to the organization. The elements of the commit-

ments are the deliverables. The cost of keeping those commitments has an estimate, as does the time required. Project teams that are well-managed and effective will meet those commitments most of the time and in a consistent fashion.

However, there can be exceptions. Product development is to some degree an unpredictable activity; there is still that estimated 15 percent unfamiliar territory mentioned earlier. If exceptions arise and commitments are not going to be met, then those exceptions should be characterized quickly and contingency plans executed expeditiously.

3. *Other operational departments are kept informed and brought on line smoothly.* The product development process involves a cross-functional team from the very beginning and is executed within an environment that includes the rest of the organization. These other parts of the organization depend on the output of the product development process and need information about its progress. It is important that these other departments receive the information they need when they need it. Have exceptions occurred? What contingency plans are being executed? The other functions then can adjust their own activities accordingly. The product is brought to production with a minimum of turbulence and waste.

4. *Design maximizes use of standard parts and processes.* It is not good enough for product development teams to make the new product work any way they can. Rather, it is incumbent upon them to make the product work, as far as possible, within the constraints imposed by the standard parts and processes that already exist within the surrounding system. To use primarily special parts or to insist on mostly special manufacturing processes is to greatly damage the potential competitiveness of the new product.

Manufacturing people and processes are key factors in the competitiveness of the organization. They must be able to build and deliver products in a competitive way. They cannot accomplish their mission if they continually must bring special processes on line to support new products. A well-run product development process must target the existing manufacturing processes to a large degree in the development activity so that when the design finally reaches a producible form, it slides smoothly into the existing processes. Manufacturing can then tune those processes and make them more and more competitive without being disrupted by the flow of new products.

There are, of course, extenuating circumstances. Sometimes a product is simply not feasible if it must incorporate all existing pre-

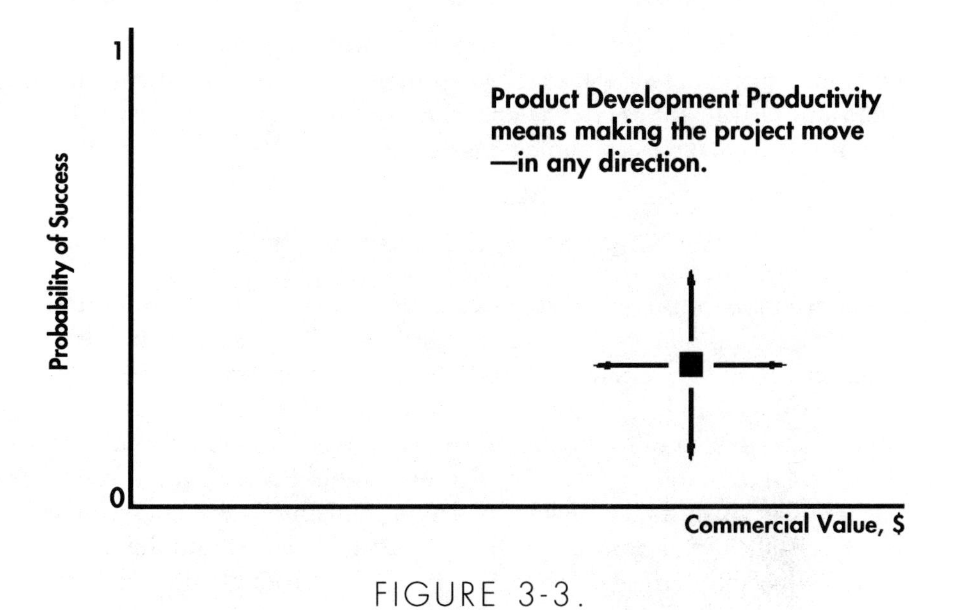

FIGURE 3-3.

An Alternative View of Desired R&D Behavior.

ferred parts and manufacturing processes. In those cases, exceptions have to be negotiated in order to make the product feasible. But a strongly desired behavior from the product development process is adherence to standard parts and processes.

5. *Product implementation is world-class and discourages competition.* This critical item refers to the discussion in Chapter 1 of the way in which the definition of a product freezes its extinction time at some indeterminate point in the future. Extinction time moves out in time only if the implementation is really world-class and discourages competition. If the implementation is less than world-class—if competitors can come in and take the basic product contribution, implement it in a more reliable, easier to use, more effective embodiment, and at a lower cost—then the introduction simply has created an opportunity for them and enabled them to reap most of the benefits of the contribution. The responsibility of the product development process, then, is to make certain that while the requirements of the product are being met, they are being designed into a world-class implementation that fits into competitive manufacturing processes.

6. *Expected return on investment is high.* A well-run product development process always will keep in mind the profitability of the effort and will foster decisions and choices that protect the return on investment to make it as high as possible.

An Alternative View

An alternative point of view is that the desired behavior for the product development process again originates with the Strategic Decisions Group. This group takes a probabilistic view of product development that locates each project on an x-y grid as shown in Figure 3-3. The vertical axis is the Probability Of Success and the horizontal axis plots the Commercial Value of the Product, should it succeed. Projects exist all over this grid, some having a high probability of success and a low payback and others being less likely to succeed but having a higher payback, and so forth. The effect of a product development process is to move this x-y position for a given project in various directions. In this view, the objective of the product development process is to move this plotted point in the grid up and to the right; but movement in *any* direction is productive.

At first glance, this approach seems a little troubling. It is easy enough to see that increasing the probability of success of the project is a constructive, productive thing to do. And it is also easy to see that increasing the commercial value of the project, should it succeed, is also productive. But to imagine that reducing the probability of success or reducing the commercial value of the project could be productive is non-intuitive, to say the least.

However, a change in perspective helps to illuminate the matter. If the product development process is viewed as a means of discovering the true nature of a possible product, the concept becomes clearer. There is a particular uncertainty at the beginning of each new product activity, and the process itself is a method for eliminating the uncertainties in a product implementation. For example, sometimes by eliminating the uncertainties, project teams find that the product simply is not feasible as conceived. Another possible outcome of the product development activity is the discovery that, while the product was initially conceived as being highly valuable in the marketplace, new information or circumstances have reduced that value to an unacceptably low level. In spite of the fact that such discoveries can

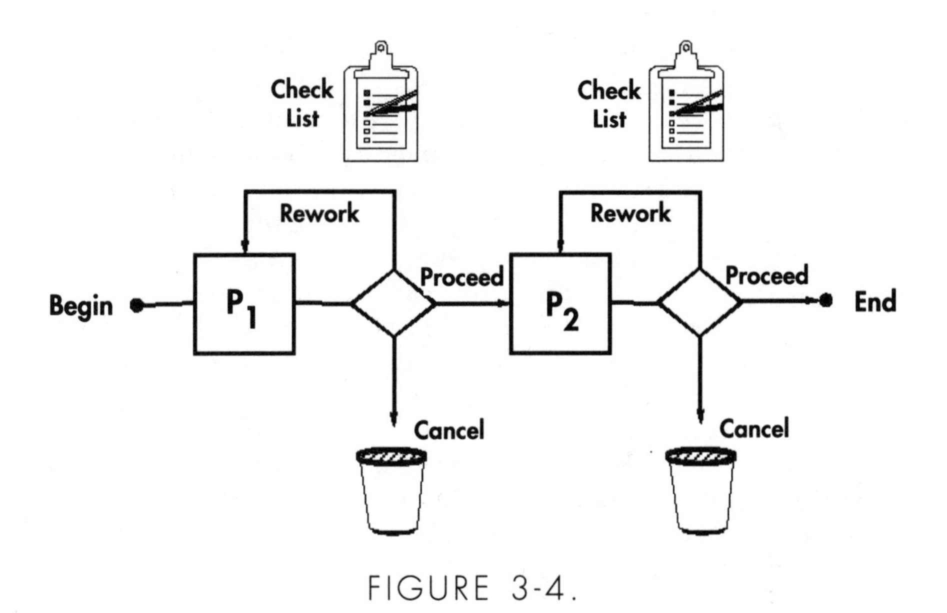

FIGURE 3-4.

A Generic Process/Gateway Model.

lead to product cancellation, both of those realizations must be considered extremely productive discoveries.

The view of the Strategic Decisions Group is that moving the plotted position of the project down and to the left, while not desirable, is still productive. To move in that direction quickly and efficiently with a minimal expenditure of engineering resources is a highly productive and meaningful execution of the product development process, insofar as that process is intended to discover what is possible and what is real within a particular product development activity. Canceling a project, then, is sometimes the most productive act that a product development team can do.

Products that are under investigation always should be viewed as cancelable. To investigate a project and discover quickly that it is not feasible for some technical or market reason is indeed a highly productive act, but it does take engineering effort and a good process. To cancel a project under investigation thus should be considered a courageous decision and should lead to rewards for the team that quickly and efficiently discovered the reality of a particular product possibility.

On the other hand, if product development teams have advanced further into the cycle—if a project has been committed to, investigated, and deemed technically feasible and marketable—to cancel after that commitment point is a more serious matter. Still, valid reasons can come up. It's extremely productive for project teams to recognize and quickly conclude, for example, that they have been outdone by the competition and that project cancellation is the only rational option. The alternative certainly is counter-productive in the extreme—to try to keep the project alive, to invest perhaps another $5 to $10 million in developing the product and to have it canceled by the marketplace. In fact, to have the customers cancel a project is the worst of all possible fates. It means a company has committed its investment but will reap no return from it.

So product development teams should be rewarded for appropriate behavior, even if it turns out that the appropriate behavior was project cancellation. And the product development process should be such that it leads to these conclusions as early as possible, if and when they are appropriate.

DEVELOPMENT PROCESS METRICS

The discussion of development process metrics is appropriate after desired behavior has been established. Consider the generic gateway model of a process illustrated in Figure 3-4. Any process can be described as a series of subprocesses separated by decision nodes leading to an end. Each decision node has several possible outcomes. One is project cancellation. Another is the decision to proceed and move on to the next subprocess. A third is to loop back and rework.

At each decision node there is a possible checklist to help make the decision and determine the outcome at that node. By articulating that checklist, development teams are well on their way to knowing how to measure the effectiveness of that subprocess. One metric of interest, particularly in product development, is how often teams have to loop back and rework what they have done. For example, if a new design for a subassembly is an input, there is a process characterized by the time it takes to generate a physical subassembly as the output. If the engineers decide that the subassembly fails to work as predicted, they send it back through the design process, effectively putting it through a rework loop. The question is, how many times did a particular subassembly go through that rework loop? The number of cycles through the loop is an important measure of the effectiveness of the basic process. If the process were perfect, the expectation would be that the first time through always would yield a successful passing of the checklist and the decision to proceed.

Cancellation frequency provides another metric for product development activities. How often do the teams get to a certain point and decide that the most effective outcome of the activity is to cancel the entire effort? What percentage of total development costs is invested in canceled projects? Too low a percentage means that projects are probably not aggressive enough. Too high a percentage may reflect poor planning or poor market understanding.

One of the problems with the generic process gateway model shown in Figure 3-4 is that it presumes that everything moves forward to a gateway where a "go-nogo" decision is made. Some projects are indeed set up that way. Nevertheless, this class of processes inherently is vulnerable to bottlenecks: generally, a lot of time is wasted waiting for the slower processes to catch up with the faster ones.

In addition to checklists, cancellations, and rework loops, teams also should look at the possibility of measuring the process itself, the P1 or P2 points shown in Figure 3-4. In general these processes have "state variables" that also are subject to metrics. State variables, as previously discussed, are those that enable prediction of the performance of the system in question.

If development teams had knowledge of the dynamic relationships in their product development process—either through analysis of historical data or through some other means—and if they knew the current state of completion, then they could predict when the process would be complete by applying process dynamics to the current measurement of progress. For example, if good historical data is available on the printed circuit assembly design process, the time to completion—once electrical designs are finished—can be estimated accurately. So the idea of state variables is an important one here. A characterization of a process by its state variables and its dynamic equations is an interesting way to understand this generic gateway model of processes.

As an alternative to Figure 3-4, consider Figure 3-5, A Concurrent Gateway Model, which provides perhaps a more realistic depiction of concurrencies in modern product development activities. In this diagram, the product development process begins with a product defini-

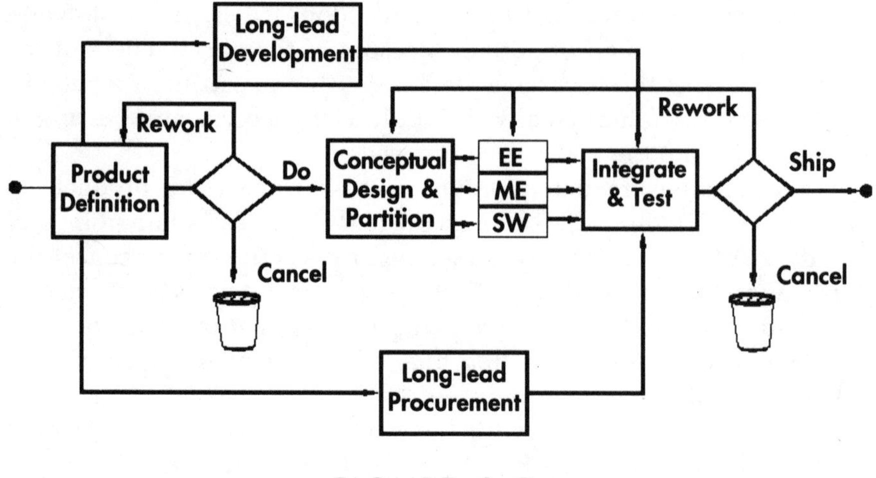

FIGURE 3-5.

Concurrent Gateway Model.

tion. Often, the long-lead items that will prevent this product from getting to market quickly become visible early in the definition phase. For example, a particular plastic molding that will take a long time to create—or perhaps an new integrated circuit—might be critical to the overall success of the project. To wait until the Investigation to Lab Prototype review is complete before initiating work on these items wastes valuable time and inserts too much delay into the process. As an alternative approach, many development teams anticipate long-lead items and launch engineering activity to develop these parts and materials at the earliest possible time. Their intent is to telescope the product development activity so that various efforts take place concurrently, and the elements come together more quickly.

Obviously, there are still some gateways involved, such as product definition and product planning. Development teams proceed through these gateways and go through the I-to-L review, all the while continuing this concurrent activity. They may even decide to rework the product definition loop a time or two. Each time they perform these checks, they also decide if anything has negated the validity of the long-lead development activities that are currently under way. Presuming that they still make sense, they continue to run.

Occasionally, as teams try to get past the gateway, they may be forced to decide that it is time to cancel this project because it no

FIGURE 3-6.

Performance-to-Schedule Commitment.

longer makes sense. At the same time, they cancel the long-lead developments. And indeed, they have created a lot of scrap. They have wasted time, energy, engineering resources, and they actually have canceled development work that someone could argue should not have been started in the first place. But the possibility of saving time is so critical that it is worth the risk of scrapping some long-lead development to achieve it. Remember from Chapter 1 that one month of saved time can bring in significant added profits, so a little time saved will pay for a lot of scrap.

If instead teams are successful in getting through the I-to-L checkpoint and they prove correct in anticipating the need for these long-lead developments, then as Figure 3-5 indicates, those long-lead items come together at the Integrate-and-Test phase of the development. They may have managed to take months out of the product development cycle. So as an alternative view to the generic process gateway model, Figure 3-5 provides a more realistic and more competitive approach to product development. It is also more in keeping with the spirit of the approach we discussed earlier with regard to risk takers.

Possible product development process metrics are divided into four categories:

1. Performance-to-commitment
2. Support for other elements in the organization
3. Post-release support
4. Engineering work force.

Taking these in turn:

1. Performance-to-commitment. As was discussed above, product development teams make certain commitments at the outset of the project. Evaluating the performance against the team's commitments provides the first series of useful metrics. Performance-to-schedule commitment is a key element in the team's overall performance. The graph in Figure 3-6 illustrates one way of representing the team's performance in this area. It displays a measure of the progress rate of a project towards its key milestones, the months of progress made per month of development activity. If the project is on schedule, that is, it is making a month's progress per month of activity, then the plot will be horizontal at the level of 1.0. If, on the other hand, a project has a milestone that is six months away and the team works six months and finds that the milestone is still three months away, then they have made three months of progress towards that milestone in six months of activity. They would be plotted at a 0.5 level on this graph.

Performance-to-schedule commitment is also a useful metric for tracking portfolio performance. The graphs can be normalized across the portfolio by simply doing a weighted average based on the number of engineers on each project. The overall performance of the product development group can then be assessed through a single graph.

Another key element of performance-to-commitment is based on the deliverables that have been agreed to early in the project. They include functionality, usability, reliability, performance, and supportability. There may be lots of other "ilities" that go as well into this list of deliverables to which a product development team may commit. One clearly important metric: what is the comparison of the actual deliverables, either delivered at the end of the project or currently active in the middle of the project, compared with the goals set at the beginning. Taking the difference and noting where deliverables have not been produced is a means of measuring performance-to-commitment. One of these deliverables might be manufacturing costs: take the difference between the actual manufacturing cost at any point in the project cycle, right to the end, and compare that to the original goal.

One of the most common reasons for project slippage is the changing of product definitions. A possible metric for this behavior might be the number of product definition changes that occur over the life of the project. Another would be the reasons for the definition changes: what factors typically trigger definition changes? Measuring the percent of canceled projects after the commitment date is essentially measuring the percent of engineering resources invested in scrap. Canceled projects should be counted as scrap if they are canceled after the commitment to develop that particular product has been made. If a project is canceled during the investigation phase, that should not be considered scrap. The reasons for either a slip in the project schedule or the cancellation of a project should be captured and analyzed for patterns. Similarly, tracking the percentage of engineering effort that ends up as scrap is useful. No scrap at all probably indicates an overly conservative operation; too high a percentage indicates a defective process.

Break-Even Time is something that can be committed to by a project team and also measured during the project activity. Before the project has broken even, a team measures not real but estimated BETs and how the estimates may vary over time. That metric is useful. It provides some insight into both the development teams' ability to estimate and the stability of the world in which they are operating. After the product has been introduced, tracking the actual Break-Even Time is extremely useful, although as was pointed out earlier, it is a seriously lagging indicator of the quality of the product development process.

2. Support for other elements in the organization. One item of interest is the effectiveness of customer input into the development process. For more or less obvious reasons, engineers on product development teams need to be in contact with customers and involved in the product marketing process. Tracking the product development function's interaction with customers is therefore of value.

Product development teams of high-tech companies also should be telling the world about the contributions they have made with their technology. Very often during the course of a product development activity, amazing technical advances will occur, but only a handful of customers will ever learn about the wonders of this innovation. Having moved on to the next project, the product development engineers often fail to document what they have done for the world. If new technical achievements have been accomplished in the process of product development activity, it is up to the engineers involved to communicate to the outside world the value of those contributions in a manner that is acceptable to the organization. The number of technical papers or articles written per engineer is one way of measuring whether or not they are doing so.

Sometimes product development activities are part of a larger system development effort. As a product development team performs its work, it must function as a member of a larger community that is developing an overall system. Measuring the percentage of overall time spent on systems-development activities as opposed to local-development activities would be worthwhile.

3. Post-release support. Obviously the percentage of warranty costs or the failure rate are ways of measuring the quality of the product that was released. The variance in the manufacturing cost against the targets that were set early in the project provides another way of deciding whether the original goals were met. After products have been introduced, the product development team's involvement in trade show activity can be extremely valuable. So the amount of time spent in that kind of activity, the amount of time spent in customer support, or the time spent in getting the product to be useful in new applications are ways of measuring the extent to which the product development team is involved in post-release support.

A key measure of post-release support is the percentage of product development time spent on continuous product engineering. That is, how much engineering time goes into a product after release either to fix bugs or enhance the product to satisfy customer needs? In this example, no engineering spent on such effort is probably too little, and 60 percent to 70 percent of the engineering effort is probably too much. There is an optimum level in between that varies as a function of the operation. But it is important to track the amount of engineering effort spent this way, because any engineering spent here is lost to new product development. Often products are introduced that require so much support that an additional cost is hidden in their development. To get caught in that trap is to sell out the future of the operation. A company introduces a line of products and suddenly finds that the subsequent rate of new product introduction has dropped almost to zero because the development team is caught up in the support problem created by the new line of products.

4. The engineering work force. The work force is a critical competitive asset for any business. The health, well-being, and competitive stature of that asset should be measured in some way. This is one of the most difficult sets of metrics to design because it requires measuring the quality of a human resource. There are, however, indicators, such as percentage of engineering time in training. Does a company really commit to and send its engineers to training classes that increase their value to the organization? The accumulated level of expertise is another. Does a company have a very experienced work force, or a naive, just-out-of-college work force? What is the overall level of their experience? The average number of years of experience is one way of measuring that characteristic, albeit not a brilliantly insightful one. Getting somewhat detailed, a company might choose to carry out periodic surveys of number of years of experience in particular disciplines.

Another interesting metric is the number of engineers a company can afford to invest compared with the competition. If a company can develop ways to measure the dollars the competition invests in a particular product development area, and then translate those dollars into numbers of engineers, it can compare the magnitude of its work force or the intensity of its effort against that of the competition.

Finally, what is the attrition by rank in the product development area? If a company ranks the performance of its engineers, and the engineers with the highest performance ranking begin to leave the company, there is a problem in terms of the professional environment of the organization. Then again, once the company has measured the attrition, it needs to learn the reasons why these top-ranked people are leaving.

These are just a few metrics that may be useful in measuring both the product development organization and the product development process.

CHAPTER 4

Side Trip: A Dip into the Meme Pool

I think that a new kind of replicator has recently emerged on this very planet. It is staring us in the face. It is still in its infancy, still drifting clumsily about in its primeval soup, but already it is achieving evolutionary change at a rate that leaves the old gene panting far behind.

RICHARD DAWKINS, *THE SELFISH GENE*

In 1976, Oxford University zoologist Richard Dawkins published a book called *The Selfish Gene* in which he presented a geneticist's view of the theory of evolution.[1] He began by reexamining the fundamentals, deriving original principles, and investigating the implications. He took as his starting point the underlying foundation for all evolution, the replicating molecule.

In the beginning of evolution was the primordial soup, an ocean filled with complex molecules, none of which knew how to replicate themselves. The earth was effectively stagnant. In Dawkins' view, in one single instant in the history of the planet, a combination of molecules came together to form a new molecule with the ability to make a copy of itself—the first replicator. Dawkins sees this first replicator as having had the potential to change the composition of the primordial soup into another stagnant condition in which its offspring dominated the chemical environment.

But evolution takes place when a mutation occurs. Occasionally, errors in replication would creep in, or cosmic radiation would cause a variation in the molecular structure of a replicated molecule. In most cases, the deviant molecules were incapable of replication, or their replicas had insufficient stamina to survive. But the mutation process is a statistical one, and the statistics were such that eventually a new molecule would appear that was more successful than the previous generation. That molecule then began replacing the previous generation with itself. And evolution commenced.

Over time, the replicating molecules became more and more complex, to the point where organisms based on gene pools developed. A collection of genes in a DNA package creates an organism that is effectively a survival mechanism for that particular gene pool. To personify them, genes have as their sole purpose replication: they strive to be more successful than any other gene pool at replicating themselves from existing materials. Dawkins' definition of a gene is a self-replicating molecular pattern that competes "selfishly" for available resources. He uses the word *selfish* even though selfishness is an attribute of thinking organisms. He claims that the behavior of these molecules is indistinguishable from selfish behavior.

As gene pools evolved, offshoots appeared that had some unique success in creating organisms that would fulfill their single purpose of replicating the founding gene pool. Each organism has one purpose in this scheme—to create new carriers of the gene pool for the next generation. All of these organisms are out there attempting to convert the world's resources into the maximum possible number of copies of their gene pools. So, in Dawkins' view, organisms are nothing more than the survival machines launched by gene pools to replicate them for the next generation.

Less than 10 million years ago, a new primordial soup developed that introduced a new kind of replicator. This time the primordial soup was a mental environment. The complexity of mental evolution, the brains of organisms, reached a state in which ideas that occurred in one brain could be replicated and passed to other brains. In a sense, the network of brains created a mental primordial soup in which an idea could create replicas of itself. Those ideas could then spread through the mental environment, consuming available mental resources.

Clearly, the requirement was a mental environment that could support an idea, presumably beginning with *Homo erectus*. The outcome was the use of fire and such tools as the stone ax. These ideas were handed down from generation to generation, examples of mental replication. Dawkins coined the word "memes" to identify these mental replicators.

A meme is a self-replicating thought pattern that competes selfishly for available mental resources. The laws that govern genes and genetic evolution appear to be completely applicable to memes as well.

One of the first principles of this particular approach is the tripartite foundation required for the growth of a meme. Replication depends on three elements: longevity of the product of replication, fecundity of the replicator, and accuracy of reproduction. That is, the meme has to live long enough to make copies of itself; it has to have a propensity to make a lot of copies; and the copies must be accurate in order to replicate the original. Those properties are equally applicable to genes and memes.

Returning to genetics, gene pools have banded together in different combinations to create different survival machines adapted to varying environments. The result is birds, snakes, humans, trees, and so forth, each of which is the result of some gene pool's ability to create an organism that propagates the gene pool.

Similarly, in the mental world, schema come together in the mind. Schema are useful, interconnected, knowledge structures that exist in the brain. Schema arise from the collection of mutually supported memes—that is, ideas that work together build a schema that acts as a survival machine for that particular set of ideas. If the schema is useful, it moves from brain to brain. It propagates itself. An individual apprehending this schema at work in one brain decides it would be useful and invests the effort to transfer that schema, including the entire set of ideas that compose it: the schema is the survival machine for the underlying meme pool.

Genetic evolution is a process that creates order out of chaos, as mutations occur that enhance the replication of successful gene patterns. Evolution flies in the face of entropy, the universal urge toward disorder. According to the Second Law of Thermodynamics, when the sun finally expires, everything will be reduced to a state of total disorder. But while the sun continues to fuel this local system, it energizes the one process that goes upstream against breakdown into chaos: evolution, where everything that develops is more beautiful and complex than that which came before.

The same is true in the mental world. Mental evolution creates increasing order out of chaos as mutations occur that enhance the replication of successful meme patterns. Some examples of evolved meme pools include cultures, religion, music and physics.

There are additional useful parallels between genetics and memetics. For example, the available gene pool for reproduction determines the quality of the offspring. Cut down the size of the gene pool below some critical level and the offspring become inbred, creating a deteriorating effect on the organism over a number of generations. Increasing the gene pool allows for improvements over generations. Similarly, the size of the available meme pool determines the robustness or richness of alternatives that can grow out of that set of ideas.

Biological organisms have developed immune systems over time. Mental systems also have done so; institutions that serve meme pools typically include mental immune systems. Dogma acts to take new ideas and throw them out as heresy. An organized religion, with its rigorous culling of its traditions to eliminate heresy, is a perfect example of a mental immune system. In science, the scientific method is a mental immune system, rejecting false memes through rigorous analysis and repeatable experimentation.

In the product development community, one occasionally pernicious mental immune system is the NIH syndrome, "Not Invented Here." NIH rejects memes from outside the local pool. In his book, *Engines of Creation*, Eric K. Drexler made an important observation on the subject:

The oldest and simplest mental immune system simply commands "believe the old, reject the new." The human body's immune system follows a similar rule. It generally accepts all the cell types present in early life and rejects new types as foreign and dangerous, such as potential cancer cells and invading bacteria. This simple reject-the-new system once worked well, yet in this era of organ transplantation it can kill. Similarly, in an era when science and technology regularly present facts that are both new and trustworthy, a rigid mental immune system becomes a dangerous handicap.[2]

NIH rejects both good and bad ideas with equal effectiveness. There are some interesting theoretical questions surrounding the reasons for the emergence of this particular immune system. The best answer seems to be that it serves to keep local processes from being disrupted. Ideally, the NIH syndrome would be altered to accept the good ideas and reject the bad ones.

In fact, good scientists and engineers should be striving to re-use ideas, to borrow schema from other domains rather than rejecting or reinventing them. Nature does it all the time. Having created a visual system, nature applied it all over the map in species as diverse as the eagle and the crab. The DNA molecule is probably the most prolific example of re-used genetic ideas: the same basic mechanism underlies all life. The respiratory system, the circulatory system, the nervous system—all passed from one domain to another, modified as necessary, and adapted into the new domain with complete success. Re-use is very respectable in the natural world; it should be just as respectable in human mental circles. And indeed, it often is. Product development teams do not customarily invent a new power supply with each new electronic product or a new transmission for each new machine. But the need to remain open to ideas from unrelated realms is just as important as re-using ideas from within familiar domains.

Now, having traveled this rather scenic back road, what is the result? Clearly, the product development process is an exercise in mental work. That being so, understanding the principles that underlie mental work will help to improve the product development process and make it more competitive. The re-use of memes and schema leads to another element in mental work that proves particularly fruitful in improving the product development process—the metaphor.

The metaphor translates memes and schema between seemingly unrelated domains. It provides a way of moving an idea structure from one domain to another. The metaphor moves comfortable, well-worn ideas into new territories, placing them in new domains where they are once again exciting and wonderful to behold.

It appears that schema that work well in one domain work well in any domain. There is an underlying structure of knowledge that contributes to the success of a particular schema. If the structure is right—if it is functional and the parts work well in relationship to one another—it creates a mental mechanism, and that mechanism is functional regardless of the environment into which it is inserted. Independent of domain, the schema is dependent only upon its own structure.

If the metaphor is well taken, and the environment that just received it is receptive to the new ideas, they work well. If the metaphor is poorly conceived, ideas imported through the metaphor do not work in the new environment and they simply break down.

Approaching the product development process with a metaphor will illuminate it from a new perspective. This metaphor will serve to make the product development process amenable to improvements so far-reaching that till now they would have seemed unlikely at best, and at worst, impossible.

NOTES

[1] Dawkins, Richard. 1989. *The Selfish Gene.* Oxford: Oxford University Press.

[2] Drexler, Eric K. 1986. *Engines of Creation; The Coming Era of Nanotechnology.* New York: Doubleday.

CHAPTER 5

Product Development: An Information Assembly Line

The principle is that if you know where you are going, having been there already, it is much better than moving from what is known towards the unknown.

EDWARD DE BONO, *I AM RIGHT, YOU ARE WRONG*

With the idea of the metaphor as a way of converting ideas from one domain to another, this concept can be applied to product development specifically. Ideally, to use a metaphor to translate ideas from one domain to another, one should find a source of ideas in a domain where they have developed more fully and proven useful. These ideas then can be moved into an immature environment where they have

Level	Characteristic	Key Problem Areas	Result
5 OPTIMIZING	Improvement fed back into process	Automation	Productivity & Quality
4 MANAGED	(quantitative) Measured process	Changing technology Problem analysis Problem prevention	
3 DEFINED	(qualitative) Process independent of individuals	Process measurement Process analysis Quantitative quality plans	
2 REPEATABLE	(intuitive) Process dependent on individuals	Training Technical practices —reviews, testing Process focus —stds, process groups	
1 INITIAL	(ad hoc/chaotic)	Project management Project planning Configuration management Software quality assurance	Risk

FIGURE 5-1.

Process Maturity Model (Software Engineering Institute).

not yet developed. So a more mature environment, a more successful environment, can serve as an example for other environments or domains that are not so successful. This search for a more mature model leads to the idea of process maturity and the Software Engineering Institute's process maturity model.

PROCESS MATURITY: A FRAMEWORK FOR IMPROVEMENT

Although the Software Engineering Institute adapted its maturity model (Figure 5-1) to assess software development processes, it seems to be useful for understanding any kind of process in any environment. SEI's maturity scale spans Levels 1 through 5, with risk of failure highest at low maturity levels[1]. Quality and productivity improve as process maturity increases.

SEI labels the beginning state of process maturity the "Initial State." The characteristics of the work going on in this state are ad hoc and in some cases even chaotic. One indicator of a process in this state is that involved individuals invent the process by which they do the work as they go along. Workers at this level start without a process in mind and then create the process and follow the process almost in the same breath. If two different people had the same assignment, they likely would invent different processes and then perform the work differently to achieve the same result. Having different processes for the same task is not necessarily a bad approach, but it is immature from the point of view of the SEI scale. The same process rarely occurs twice at the Level 1 stage. Everyone will be inventing something slightly different, and there is no memory in the system.

Over time, a single process begins to evolve as the winner, somewhat like the "survival of the fittest." A "Repeatable Process," Level 2, is characterized as an intuitive process because it is dependent on the individuals involved. The individuals remember how to perform the process, and they simply continue to repeat that successful pattern.

Once proven successful, the process eventually is written down. Effective documentation moves the process up to Level 3, the "Defined Process," where it is at last independent of individuals. If the process document is created on the West Coast, it could be shipped to the East Coast where someone could read it and replicate the process. However, if the document is simply sitting on someone's shelf gathering dust and the process is still being controlled by the individuals involved, the process doesn't advance to Level 3. Level 3 implies that documents carry the process and that people refer to the documents to keep the process on track. At Level 3, documents are working documents.

Eventually, the documented process moves into a new stage. At this stage, development teams define measures that represent the important states of the process and then begin to track those measurements. This is Level 4 or the "Managed Process." Teams use the measures that they have developed to determine whether the process is on track, and if it slips off track they put it back where it belongs with corrective action.

At the "Optimizing" level of maturity, Level 5, the process is no longer constant over time. Instead, it is actually in a state of continuous improvement. Not only are teams measuring the process, they can use the results of those measurements to determine how to improve the process, make the changes to the process and use the measures to verify that the changes have indeed made an improvement. That is an ongoing improvement cycle.

In most cases, people probably can apply this scale intuitively by looking for characteristics that reveal the maturity level of the process in question. At Level 1 people will invent the process as they go along. If they are asked what step they will move to after they have finished the current one, they reply, "I really don't know. I know I need this next piece; I'm not sure how I'm going to get it, but I'm going to try this next approach and see if it gives me what I need." However, if they can tell us what the processes are or what the activities are ahead of them, then they are probably at Level 2, because they have been through this process before. They know what they are going to do next, but they are not following a document. They are simply working from memory.

If, though, they have a document they can carry across the country, hand to a qualified group, ask the group to build their widgets on the basis of the document, and the new group can do it, that's a Level 3 process. If an information system exists to keep the process on track—if data entry is done at various points in the process, and there are reports coming back—then the process is at Level 4. The existence of a computer information system tracking the appropriate state variables for the process definitely indicates a Level 4.

Finally, a Level 5 process exists when someone in the organization is responsible for ongoing improvement of the process and has established improvement goals and measurement systems. Furthermore, this individual tracks progress toward those goals and keeps the improvement program on track.

THE INFORMATION ASSEMBLY LINE METAPHOR

As suggested above, to accelerate process improvement, look to a more mature world and map the ideas from that world into a less mature world. The metaphor that will apply here does that; it maps memes and schema from the manufacturing domain back into the product development domain.

An educated guess of the maturity levels of various processes in both the product development domain and the manufacturing domain is shown in Figure 5-2. In manufacturing, the large majority of processes are bunched at Level 4, with a few at Level 3 that are docu-

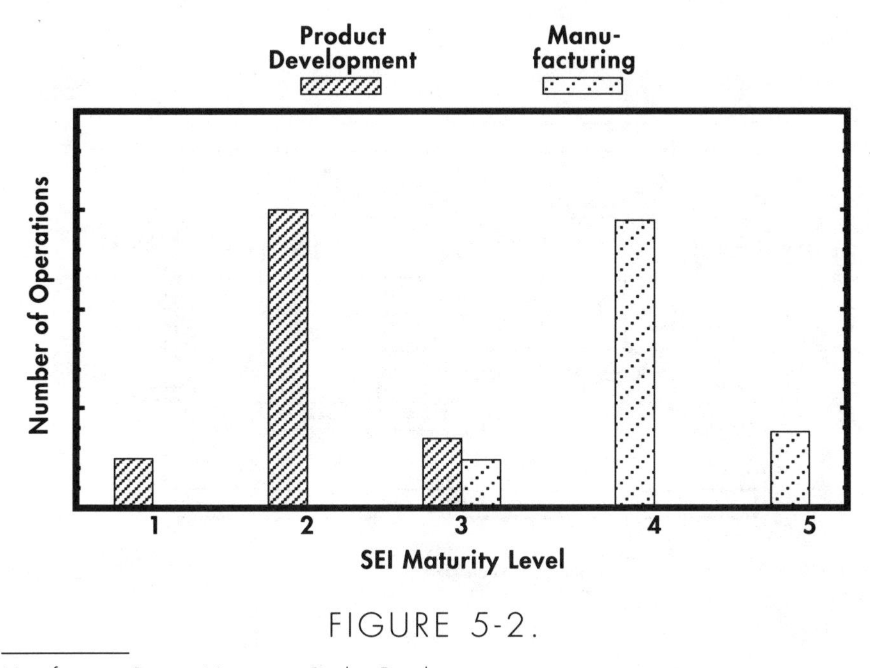

FIGURE 5-2.

Manufacturing Process Maturity vs. Product Development

mented but not measured, and a few at Level 5 that are being improved continuously. The representative histogram for product development reflects that of manufacturing, but shifted down two levels. Most product development activities are repeatable; the teams have been through them before. The activities have a life cycle. They may even have been documented, but the actual day-to-day activities of engineers and the processes that engineers go through are much less well documented. In product development, engineers are working primarily to remembered and perhaps agreed upon processes.

With the metaphor it's possible to map ideas that have proven successful in the more mature manufacturing domain and apply them in the product development domain. The metaphor is one that compares the factory assembly line to the product development process.

The Japanese have raised the management of factory assembly lines to a Level 5 state of maturity in many cases. They are teaching many companies about just-in-time manufacturing and time-based competition, and they have been continually improving the assembly line and the factory as an institution for the last thirty years. The ideas that work—that serve to move the state of maturity of the manufactur-

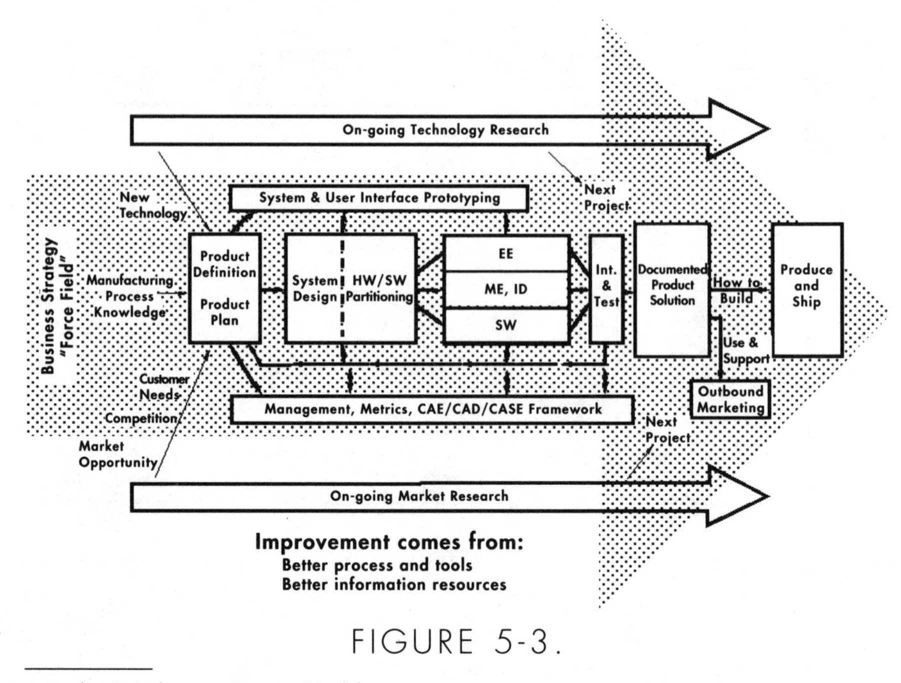

FIGURE 5-3.

A Product Development Process Model.

ing assembly line forward—map precisely through the metaphor back into the product development domain.

However, the metaphor first must be tailored to the purposes of product development. The assembly line is a mechanism or a process for adding value to parts as rapidly as possible at each step until they become a quality finished product ready to ship. Parts and materials come into the assembly in a raw, rough, less valuable state. The added value changes the state of the parts and materials. With enough added value, they can be shipped to customers who will pay for them. That is, the assembly line has produced a finished product.

W. Edwards Deming and others have come up with the improvement paradigm for assembly lines. Deming begins with the need to measure and stabilize the process first, and then improve process productivity and quality once the process is under control. One element of the Deming approach requires improvement in the quality and timeliness of input materials. Another impels elimination of waste and variation in the output of the process. Once these elements are under control, the response time of the assembly line can be shortened. That

Product Manufacturing =
Procurement, fabrication, assembly, and shipment of parts

Transforms into:

Product Development =
Gathering, creation, integration, and documentation of information

FIGURE 5-4.

Transformation of Concepts from Manufacturing to Product Development.

is, the process can speed up. However, if companies start to speed up the assembly line prematurely before they have measured and stabilized it, they create chaos. If they speed up the assembly line before they have both the productivity and quality of the process well under control, they create scrap.

An equivalent look at the product development process views it as an information assembly line. It is a process for adding value to the information set as rapidly as possible until that information set describes the manufacture, support and use of a quality new product. Again the process begins with raw, rough information. Each step adds value to it.

Figure 5-3 depicts the product development process. Fueled by enthusiasm for the business strategy, cross-functional development teams take rough information about customer needs, technology and market opportunities and turn it into an exact definition of the desired product. At this point they already have added a great deal of value to the information. The remainder of the product development process puts the information through a series of additional value-adding steps, to the point where the information set finally describes how to build, support, and use a brand new product.

If adding value to information in the product development process is approached as an assembly line, then the improvement paradigm should be the same as that for a manufacturing assembly line: first, measure and stabilize the process. Measurement and stabilization are not trivial problems, particularly for development teams who have not been accustomed to measuring R&D or product development processes in the past.

Once team members learn to measure and stabilize the process, they then can begin to define and improve its productivity and quality. Having defined the quality of product development information—an interesting challenge in itself—they are ready to improve the quality and the timeliness of this information. They also can tackle waste at this point. (More about waste control will be considered in Chapter 7.) And finally, if they bring all these factors under control, they can shorten the innovation cycle time. Shortening innovation cycle time is equivalent to decreasing the response time of an assembly line. The time-based factory has an equivalent in product development.

Figure 5-4 describes product manufacturing as the procurement, fabrication, assembly, and shipment of parts. The metaphor transforms that description word-for-word into the accumulation, creation, integration, and documentation of information.

In the light of this metaphor, the equivalency is startling. Each word in the manufacturing domain has its equivalent in the product development process. Procurement is equivalent to the gathering of information. Fabrication of materials is equivalent to the creation of information. Assembly of parts and materials is equivalent to the integration of information. Shipment of parts and materials—or a finished product—is equivalent to the documentation of information. The parts that flow through an assembly line are equivalent to the information that flows through the product development process. Each of these concepts is equivalent; one converts into the other.

The metaphor works. It transforms a myriad of memes and schema exactly through these relationships. Figure 5-3, the Product Development Process Model, is effectively a model of the product development assembly line. Every block in the diagram is an information process, while everything that flows down an arrow is information. The process begins on the left with rough, raw information coming in. That information includes new technology, manufacturing process knowledge, and rough information about customer needs, competition, and market opportunities. The product definition process block can be documented and indeed has been documented successfully within Hewlett-Packard Company. Having documented it, HP is, of course, working to improve it.[2]

The first process block takes raw material coming in, in this case raw information, and transforms it into a very precise description of a product, thereby greatly increasing the value of the incoming information. That information flows into the next block, the system design block, in which a system is designed to create a new product. Part of that system design requires the work be partitioned into a set of requirements for electrical engineers, mechanical engineers, and software engineers. System design and work partitioning are, indeed, other value-adding processes.

Throughout the product development process there are feedback loops. At the system design stage, for example, team members might consider doing some interface prototyping. They create a user interface design and try it back in the Product Definition block with the users. Likely users are asked, "Is this what you wanted? Does this prototyped interface represent a product that would be a solution to your need?" Team members use the feedback from potential customers to generate a more tightly verified product definition.

The project team members then follow up with another, more focused system design until, through these iterations, they get convergence. Eventually the output of that system design is a set of requirements for the electrical engineers, mechanical engineers, and software engineers. Presumably, these are solid requirements that reflect the optimum partitioning and architecture for the system, and that create exactly the right product to address the given opportunity. No one is wasting time designing the wrong products or the wrong features.

Members of each group then take the design requirements and turn them into schematics, drawings, or source code that represent an information transformation. This level is beyond invention and definitely into process. The electrical engineers, for example, will design filters, amplifiers, microprocessor systems, memories with given characteristics, and so forth, all of which are essentially straightforward processes.

The output of each of those design disciplines is then integrated and tested as the first prototype product. At this point the loops are invoked again; measured performance is fed back and checked against design requirements. As a result of test failures, the team may have to redesign some product sub-assemblies or components.

As before, Integration and Test is a value-adding information process. This block takes information representing designs and integrates the designs into a prototype that is used for functional tests. The inputs to this block are designs that have not been integrated or tested yet; the output is an integrated design that has been verified. Clearly this process adds value to the information set. Finally, the documentation step collects all important information about the new product and transforms it into usable form. From there it is passed along to where it will be used best, in manufacturing and marketing,

The layout of Figure 5-3 and the above discussion imply an unintended time sequence to these activities. Many of the activities in these process elements can and should be concurrent. For example, product documentation should begin early in the development process and proceed concurrently with product design. Too often, documentation is not thought of until the very end and then written in a hurried and haphazard manner. This results in poor quality—and perhaps lost or neglected information—in the essential output of the process.

Often, long-lead parts can be identified as early as the product definition step. It is usually appropriate to begin the design of these parts as early as possible even though a significant cost in scrap might be implied. This cost is usually trivial compared to the value of the potential time savings achieved through concurrent activities.

Figure 5-3 then is a fair representation of an information assembly line. The product development process sits embedded in an environment that is created by ongoing market research on one side and ongoing technology research on the other, both taking their direction from the energy field provided by the business strategy. Ongoing research triggers the beginning of this particular instance of the product generation process. At some time in the future, when another project team becomes available, it triggers another project activity, and then another and another, whenever the resources become available.

In a well-designed organization, research in both technologies and markets will be ongoing, background activities that are concurrent with the development of specific products. This work is symbolized by the long arrows at the top and bottom of Figure 5-3. The ultimate purpose of these background activities is to identify new opportunities for innovation as early as possible so that development teams can attack the highest priority opportunity as soon as their current projects are complete. Moving quickly from a completed project to a new one minimizes the empty time between opportunity and project activity discussed in Chapter 1 and maximizes the productivity of development teams. In many organizations, time lost in moving a project team from one project to the next is a major source of waste and lost profits.

Intelligent product development, however, can work only within the context of a rational strategic framework. In addition to the background activities mentioned above—ongoing market and technology research—perhaps the most important background activity is the development and continuing refinement of an abundantly clear product strategy. It is critically important to have a business strategy in place into whose framework each new product fits clearly and cleanly.

As with assembly line advances, improvements in the product development process model come from two sources: better processes and better inputs to the processes. To improve an assembly line, companies need to look at their processes and ask where the processes are creating flaws in the products. Then they work on those processes. They look at the tools they are using and ask, "Where do these tools create flaws? How do they slow things down, and how can we improve our productivity with better tools?" And then they design their tools. They perform this analysis for every workstation on an assembly line until they figure out how to optimize it.

In addition to improved processes and tools, improved quality in the output depends on improving the input, that is, reducing the flaws in the incoming materials. Similarly with the product development process; To improve it, companies must improve the processes within it, the tools they use and the incoming information resources. This is simply a transformation of the Deming approach.

Deming is pertinent here. As in manufacturing, the first step in improving product development is recognizing that it is a process. This observation is in no way intended to trivialize either the individuality or the creativity of the participants in the product development process. Nevertheless, establishing this connection—assimilating this meme—is a powerful first step toward improving the product development activity.

Deming's next step is to stabilize and measure the process. Once companies have done this, they can control the quality and the productivity of the process.

The quality of the output has to do with how well the information being created at any step in that process meets customer needs. How well does the information describe the product? How free of errors is it? Productivity measures the amount of investment required for a given amount of value added to the information set. If companies can achieve the same result with two fewer engineers, that becomes an important improvement in productivity. Alternatively, if they can get the same result with the same number of engineers in half the time, that is also an improvement in productivity. So, to the extent that companies can measure the information coming out of the process—again, no trivial task—they can begin to talk about the productivity of the process.

Finally, when they have improved quality and productivity, and they have the process under control, they can begin to remove time. This is the most critical competitive step. It is the carrot, the pot of gold at the end of the rainbow. It is the reason for investing huge amounts of resources in understanding the process. Until companies can understand, measure and control the process, they will find that trying to shorten it will result in scrap and chaos. There are no shortcuts. They can ask their engineers to work faster; they can ask them to work longer, and yet they won't achieve the results they want.

Alternatively companies can explore the process—how to make it faster, how to make it more efficient, how to make more things happen per unit time, and how to engineer concurrently as opposed to sequentially. In so doing, they ultimately will follow in the footsteps of the people who have explored corresponding paths for years in the assembly line environment in order to take time out of that cycle. They will find the bottlenecks and try to remove them by adding capacity or by putting in concurrency to create parallel paths around them.

One other aspect important to the product development process relates to the quality of the incoming information. Garbage in/garbage out is just as pertinent here as in any other realm of computer data processing. To improve information resources available to the engineers, companies need to look at the timeliness, the accuracy, and the usability of that information.

Of these three considerations, usability may be the most problematic. Engineers are buried in information. The nuggets are there but not really usable. Although the information may have come in on time and was available on their desks, it was unreasonably obscured. Ideally, companies would create an environment for engineers where the information they need is as easy to get at as turning around and reaching. This convenience is analogous to a worker on an assembly line reaching around for a new box of parts when the previous box is exhausted. Just-in-time information delivery will work just as well for engineers as just-in-time material delivery works on the assembly line.

The negative is equally true. If the engineer turns around and the information is not there, if the assembly line worker turns around and the required parts are not there, and if they have to go search for what they need, they have both wasted time. If engineers have to invent a way to find the information, then they are not spending their time developing or adding value to information about the product. They are wasting time looking for the information they need. In the worst case, they decide they do not need the information. They forge ahead without it and create the wrong product, or a product that does not meet critical requirements, because the information was not there when they needed it.

Consider, for example, a printer designed for global distribution. If the engineers developing the printer do not know or have information about character sets for the various alphabets around the world, from Eastern Europe to Asia, the chances are they will design the product so that it supports English only. When the time comes to localize the product for Korea, they will have a serious problem because they have to re-engineer the whole thing. Every month spent re-engineering the product is a month by which the market window in Korea is diminished. Multiply this problem perhaps fifteen times around the world—for Eastern Europe, Western Europe, Asia, China, Japan, Singapore, etc.—to realize the extent of the market being lost. Whereas, if the engineers know from day one that the product requires any number of different sets, they can obviate that re-engineering process by picking the right architecture at the outset.

Another concept that transforms well from manufacturing to product development is "cost of quality." The cost of an error in quality depends on when the error occurs in the assembly process and when it is detected. If the error is created in the very first step in the line and is later detected by the customer, it can be terrifically expensive. Manufacturers have created a flawed product very early; they have added value, built parts on top of this error in quality and added more value to these parts. The early error may cause the product to be totally worthless in the end. If the customer discovers this error, an additional element of the cost is the customer's dissatisfaction. Manufacturing companies may be able to fix such errors, but if they are buried deep in the heart of the products and have had entire products assembled around them, it is going to be extremely tough to implement fixes. They may end up having to discard the products and give the customers new ones. So the cost can be enormous if an error is created early and discovered late. On the other hand if companies make errors in an assembly step and detect them at the end of those steps—in time to go back and rework the products—the costs are then small by comparison. Obviously, the best approach is to make no errors at all.

All of the steps above for an assembly line are absolutely similar for product development. Referring again to Figure 5-3, if development team members make an error in product definition and go through all the work downstream that is implied in the development cycle, by the time they are done they may have invested perhaps millions of dollars in creating the product, which in this case is an information set that describes a new system. If the customer then tells them that they have failed because they omitted a particular feature or requirement or seriously overpriced the product, the cost of their error is immense in terms of customer dissatisfaction and wasted effort.

If, on the other hand, development team members define a product and incorporate a flaw but put it through a feedback loop—let us say they prototype the user interface and show it to some users as part of their check cycle—they have a chance to fix the product definition before they have built failure into the system. They fix the product definition; they come back through another verification loop, and perhaps this time they see some excitement in the customer's response. By comparison the cost is very small. They made the same mistake but they caught it, fixed it in a very short closed loop, and got on with adding value to the right information. These steps are equivalent to the manufacturing assembly line example. Cost of quality is exactly the same. Once again, the concepts map.

Looking at it another way, the quality of the incoming materials is equivalent to the quality of the incoming information. This includes good information on new technologies, good information on world-class manufacturing processes, and good information about customer needs, competition, and market opportunity. The better the quality of that incoming information can be made by the development teams, the better their chances of producing a successful product.

Still another concept that maps into the product development process from manufacturing is the Deming dictum that quality must be built in, not tested in. To test in quality is an extremely expensive approach. The most expensive way to achieve quality is to produce a junky product, test it, find out that it is bad, send it back, rework it, and continually revisit that loop until it finally passes the test and ships. The way to attain quality is to ascertain that every step in the process is a quality step; the end result is then a quality product. The product is tested once and shipped. Ideally, companies become so good that they need not test at all because it is not worth their time.

The same goal applies to product development: make the product development process so good that the product never needs to be tested. Building quality in at each step is the ideal. Product development cycles inevitably have a test-in-the-quality philosophy. Building quality in before testing—using quality incoming information and quality information processes so that a perfect result is created every time—is not a philosophy that has attained widespread prevalence in the product development domain. What companies normally do in product development is create the design and integrate it. Once they have a product, they test it to see if it does what it is supposed to do. Then they build a loop out of that test to go back and rework whatever is needed. But the more they build in quality at the front end, the less important the loops become. The loops may still be there, but they can certainly be minimized—and the implied costs minimized along with them.

When development teams hit the Integrate and Test function and discover that they are miles off the target product, their products miss the market window and costs begin to skyrocket. Then the product must go back through the system to be reworked, big pieces of the design scrapped, most of the assemblies redesigned, and an unscheduled six to nine months is added to the cycle, diminishing the market window by an equal amount. That is the cost of poor quality. Clearly, building quality in rather than testing it in offers an enormous opportunity for improvement in the product development domain.

LIMITS TO PROCESS MATURITY IN PRODUCT DEVELOPMENT

How far can product development shift up the maturity scale? A great deal depends on the circumstances of the market, the turbulence of the particular environment, the changing state of the technology, and many other factors. If they are all unfavorable, a company might be able to get to a Level 3 at best. But if companies are doing something relatively straightforward, where each new development cycle is similar to the last, they might get to a Level 4 and even conceivably to a Level 5.

For example, if an organization's mission is to ship power supplies designed to individual customer specifications, it is conceivable that the processes could get to a Level 5. The science and technology that go into power supplies are probably stable enough—transformer technology, switcher technology, and some application specific integrated circuits to control operation—that engineers can take the same set of specifications, move the numbers around and come up with a design that meets the requirements. They could identify the key points in that process from input specifications to output design, track performance all along the way, and conceivably even put in place a continuous process improvement scheme that would put them at a Level 5. It is important to note that Level 5 does not imply automation, but it is certainly more amenable to automation. Nor does it imply a lack of human involvement. Level 5 does, however, imply a superior understanding of the process, a knowledge of the key state variables, and the ability to apply continuous improvement approaches to the process.

In general terms, the productivity and quality of the process go up as the maturity level goes up. The risk associated with the outcome of the process goes down. While these results generally are true, there are some interesting caveats about process maturity.

Each organization probably has an ideal maturity level on the SEI scale. The maturity varies as a function of the nature of the organization. If the organization in question is a manufacturing shop whose job is to turn out crankshafts, its objective is to get cost down and increase the work flow through that process. That organization needs at least a Level 4 and preferably a Level 5 process.

On the other hand, is a Level 4 process desirable if the organization is a start-up company working to develop a new product with new technology and deliver it to a new marketplace? Should that organization hire an R&D manager with an intimate knowledge of the world's best, most sophisticated product development measurement system? Should it put in place a process that has an information system to track all of the states of the process with computer regularity and to present weekly reports on the process? The chances are that person is entirely inappropriate for a start-up company. The start-up wants a very small group of highly motivated engineers who can invent whatever process they need in order to get work done. They have to be down at Level 1. A Level 4 manager in this situation would probably demolish the company. In short, Level 1 is not necessarily bad if the circumstances are appropriate. So there is an evolution of businesses that begins with

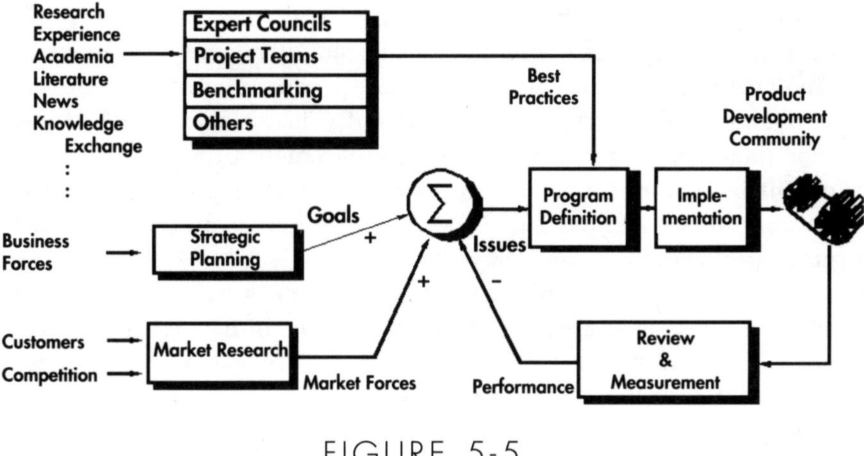

FIGURE 5-5.

A Continuous Process Improvement System for Product Development.

the start-up at Level 1, and as its product development matures, it moves up to a Level 2 or 3 and beyond in some cases.

A common occurrence in turbulent businesses is the disruption of the product development process by frequent reorganization. Labs that have achieved a repeatable Level 2 development process depend on the same key individuals being involved with each new product iteration. When these labs are split up through reorganization, maturity of the development process often reverts to a Level 1 as development teams are divided and separated from familiar support systems. Process improvement activities in such situations are generally futile, being completely negated by organizational disruption of the targeted process. If process improvement efforts can be futile, when can efforts to improve product development succeed, and when are they simply a waste of energy?

Insight into the criteria for success of improvement efforts can be gained by applying control system principles to Figure 5-5. Continuous process improvement is subject to some of the same laws that govern dynamic control systems. Readers with a control engineering background will recognize this diagram as a feedback control loop. The product development process is the element that needs to move in a controlled way and in a direction that makes it more competitive.

The success of such efforts depends on the relative stabilities of the process being improved, the forces that influence it, and the organization that supports it. Improvement success also depends upon how quickly people can measure results, identify competitive issues, define corrective action, and implement process change. To succeed, improvement efforts must be significantly faster than other change mechanisms that are inherent to the development process or the environment in which it exists. If improvement efforts are too slow, they may start right but turn out to be ineffective or even detrimental to process performance when they finally take effect.

An organization's ability to move its processes up the maturity scale increases as the speed of measurement and corrective action exceeds the inherent rate of change of those processes and the forces that affect them. The converse is also true.

NOTES

[1] Paulk, M.C.; Curtis, B.; Chrissis, M.B.; et al. *Capability Maturity Model for Software* (CMU/SEI-91-TR-24, ADA240603). Pittsburgh, PA: Software Engineering Institute, Carnegie Mellon University, August 1991.

[2] Wilson, Edith: "Product Definition: Assorted Techniques and Their Marketplace Impact," Proceedings of the 1990 IEEE International Engineering Management Conference, pp 64-69.

CHAPTER 6

Adding Value to Information

As work becomes increasingly information intensive, organizational success will depend more and more on giving each individual contributor needed information at the right place at the right time and in the right form.

ARNO PENZIAS, "IDEAS AND INFORMATION," *MANAGING IN A HIGH TECH WORLD*

The purpose of the product development process is to add value to information in a systematic way. But what does it mean "to add value to information?" An early pioneer in information transmission technology, R. V. L. Hartley, provides part of the answer. His equations show that the value of information carried by an event is inversely related to the probability of the event. Adding value to information thus becomes, to some extent, an exercise in designing experiments and in working to move probabilities toward certainty. But the manufacturing metaphor reveals that there is more to it.

To understand fully how value can be added to information, it is also necessary to examine the various factors that contribute value to parts and materials. This probing yields important philosophical underpinnings for the time-based product development cycle.

A CASE STUDY—THE STANFORD CLOCK

A tower on the Stanford University campus in California provides a fitting home for a wonderful old clock. At the top of this tower four iron clock faces with Roman numerals display time to the cardinal points of the compass. Bells in the tower signal the time to those who are beyond view. Every fifteen minutes the chimes ring out, singing progressively the familiar "Big Ben" theme. And of course, on the hour, a deeper tone with a more stately tempo tells all who can hear just where they are in their day.

Windows in the base of the tower allow those with idle time to observe the mechanical innards at work. Benches are provided as an invitation to linger. One can watch at close hand the pendulum swinging to and fro, gears meshing, pulley rigs pulling.

The clock itself is a mechanical assemblage installed in a robust steel frame. The parts are familiar to those who understand clocks: a heavy pendulum connected to an escapement mechanism that, in turn, moves a gear train. The gear train rotates a vertical shaft that disappears into the lofty heights of the tower to turn the clock hands. Power is provided by a winch connected by cable to a substantial weight.

More interesting, perhaps, are the separate mechanisms that ring the chimes. At proper intervals the clock mechanism trips a latch that allows another winch to turn a set of cams. The speed of this action is set by large brass vanes that provide air resistance as the machine turns. Each rotating cam lifts and drops a push rod that rings its assigned chime. Once the sequence is complete, the latch is reset and the action stops. Cables attached to weights again provide power.

Some History

As the story goes, this clock originally was installed in a church tower on the Stanford campus in 1901. In 1906 the tower was destroyed during the San Francisco earthquake. The clock was salvaged and put into storage. It was restored to service in 1913 in the reconstructed church where it performed its function for many years. In 1967 the church tower was torn down and the clock was put into storage once again. There the mechanism gathered dust until 1980 when an interested mechanical engineering professor resurrected the old clock and

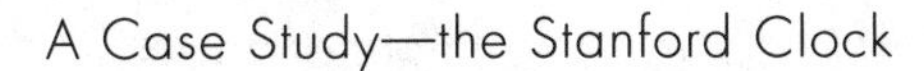

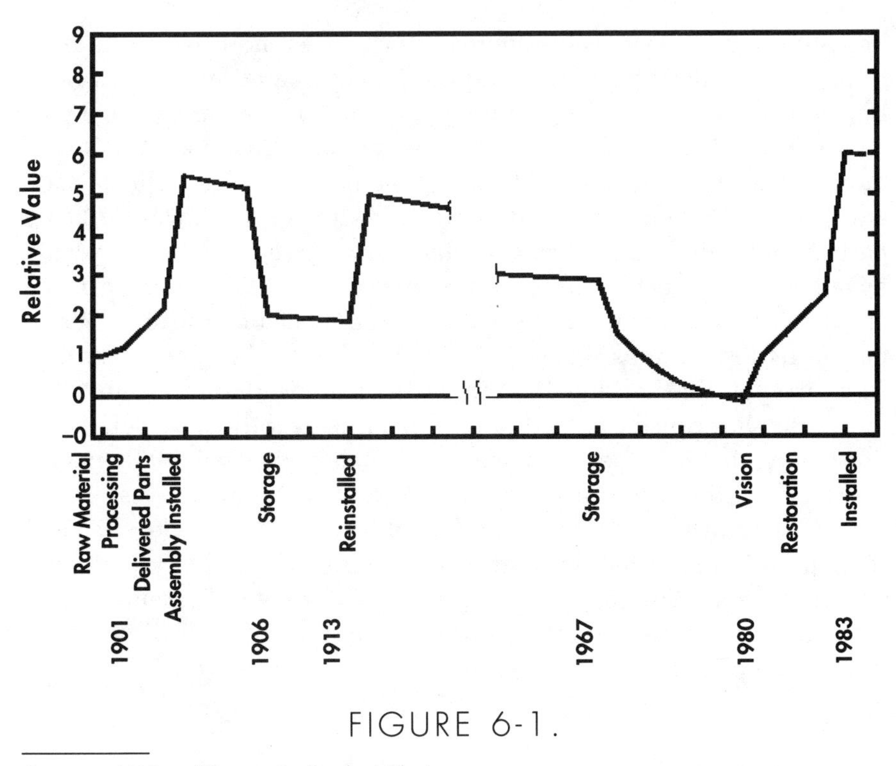

FIGURE 6-1.

Conceptual Value of Parts in the Stanford Clock.

got it working again. The current tower was built in 1983 especially to house this marvelous timepiece and its bells.

The Changing Value of Clock Parts

When assembled and working in a public place, the Stanford clock provides a valuable function to many people. Disassembled and scattered about, however, the individual parts of this clock lose their value. During the long time that they were in storage one can easily imagine them stuck away in some dark place gathering dust. Many of us have similar parts and materials cluttering our own garage, underfoot and often useless. The worth of this stuff—some would call it junk—is often less than zero, since it requires an investment of effort just to haul it away.

Figure 6-1 shows a conceptual assessment of the relative value of each part in this clock over time. As a part began its life as raw material, its value was modest. Value increased as it was modified to its final form, perhaps through casting and machining operations. It grew even higher as the parts were brought together for final assembly. Each part attained peak value when the clock was finally installed and set in motion. During the times that the clock sat in storage, the value of each part ebbed and then continued to decline with age. In the late '70s the parts conceivably could have reached a negative value as they may have been considered old junk to be discarded. When interest in the clock rekindled, however, its value rebounded and ultimately surpassed its former apex when the clock was restored to public service. As an irreplacable antique, it is more valuable now than it was in 1901.

What has caused this difference in value over time, especially for those parts that have hardly changed since they were originally made? What has restored worth to these parts and materials after so many years? To answer these rather philosophical questions is to discover some underlying principles that determine the value of physical parts and materials. With those in hand, the information assembly line metaphor gives insight into similar principles at work in product innovation processes.

FACTORS THAT CONTRIBUTE TO VALUE

Opportunity

For the Stanford clock and its individual parts to have any value at all, an opportunity must exist that they satisfy. Chapter 1 defined an opportunity as the coexistence of customer need and an appropriate technology. The opportunity for the clock can be elaborated and described in terms of coexistent value to the user, value to the customer and feasibility of implementation. The user is anyone who needs to know the time or appreciates the aesthetic value of the clock. The customer is the individual or institution that is willing to spend money on the clock. Feasibility of implementation includes consideration of cost, logistics, and technical matters.

All of these factors must be favorable before an opportunity exists. Once these conditions are met, the value of the clock and of its independent parts is, at least conceptually, proportional to the product of these three factors. If potential users attach a high value to the aesthetics of the old clock and the service that it provides, if a philanthropist gains great satisfaction in providing a tower to house the clock, and if the restoration is a straightforward task that is not too costly, then the parts of the clock have great value. If any one of these three factors fails, the clock loses its worth.

A Compelling Vision

In 1980, a Stanford mechanical engineering professor decided that, for historical and aesthetic reasons, the old clock should again grace the campus. He began restoring the old parts and he persuaded a philanthropist to donate the funds for the tower. After three years of dedication, his vision was rewarded: He was able to return the clock to its rightful place at the center of Stanford consciousness, a visible and aural link with the proud traditions of the past.

When the old clock was removed from service in 1967, its value ebbed to an all-time low. It was most likely considered junk during this period. When the professor first envisioned the clock restored and once again pealing out the time to Stanford students and faculty, this venerable mechanism began to regain its value, even though it was still gathering dust in some dark basement. The professor's enthusiasm provided the motive force that propelled events until the clock was once again restored to service in its new home. Without a champion such as this, the value of the clock would have remained low, and it undoubtedly would have come to a much less glorious end.

Existing Intellectual Framework

If no one had invented the pendulum-driven escapement mechanism, this collection of parts would be without meaning or value. No one would think to assemble them in this unique manner and set them in motion keeping time. The series of inventions upon which this device depends have created a body of intellectual property that provides an essential ingredient to the value of each part in the mechanism.

Processes that Increase Functionality

Each part in the clock mechanism started out as raw material. A gear, for example, began as iron ore, was smelted into an ingot, then cast into rough form, and finally machined and polished into an elegantly functional form. Each process step added value until a smoothly working part was created that fit perfectly into its singular place in the clock. The techniques applied in fabricating this gear were, in many cases, well established and standard in the manufacture of a wide variety of items. Others may have been uniquely developed to build parts for this particular model of mechanical clock. In either case, the outcome of each action was well characterized; the success of each subsequent operation was nearly certain. This well-defined process is, of course, a fundamental strength of mass production.

Congruence of Place, Time, and Need

As the parts for this clock came together from their various points of origin to the place where they were assembled, their value increased. Parts are more valuable collected together at the place where they can be put to use than they are when isolated on a shelf in a warehouse. Furthermore, the parts must be brought together in time to fill the need. If they arrive late and another solution has to be found, then the potential value that they might have provided is lost.

Usability

To have value, parts for the clock had to arrive in a form that could be used. Gears had to have been made for the proper shaft size; screws must have had the proper dimensions and thread pitch. The parts had to have been compatible with any assembly tooling that was used. A part that works in every other way but that requires an unavailable tool for proper installation has no value.

EQUIVALENT FACTORS IN PRODUCT INNOVATION

Opportunity

Corresponding to the opportunity addressed by the clock are market opportunities addressed by a new product. Market opportunity includes more than just need, however. It includes potential profit, feasible implementation, and the chance of a favorable competitive position as well. Opportunity increases with stronger demand for the product, greater profit potential, greater feasibility, and more favorable competitive position. As opportunity grows, the information set that describes the new product increases in value, and the efforts to build that information set become more consequential.

Conversely, if the opportunity is low, the same product development activity can have little or even negative value. Far too many excellent product development efforts fail to yield a profit because they were not well aimed at an appropriate market opportunity. For example, product innovation activity is too often launched by an exciting new technology—one that has compelling appeal to the engineering team. But technology is secondary to the elements of opportunity. If it increases the profit potential or customer demand, or if it improves competitive position, then a new technology contributes to the opportunity. If it does none of these things, it can be a terrible distraction and a waste of time, money, and human effort.

A Compelling Vision

In the domain of product development, product champions correspond to the professor in the Stanford clock story. People who perceive an exciting opportunity and describe it enthusiastically and repeatedly to everyone who will listen can play a pivotal role in realizing a new product. Effective product champions motivate the project team simply to swarm obstacles that get in the way of success. They convince upper management to invest scarce resources in risky ventures and get customers excited about new possibilities. They invent processes and do whatever becomes necessary to bring the vision to a reality. The clarity and excitement that these individuals provide can multiply the value of the work done by the project team. They focus the efforts and keep development teams on track toward a valuable goal.

Vision such as this is like a spotlight, searching out information that can help realize a particular product dream. Like the clock parts gathering dust in a garage, information can go unnoticed and have little apparent value. However, when highlighted by someone's vision as an essential ingredient in a product solution, this very same information suddenly begins to take on new value. Insight into how a particular bit of information contributes to a solution thus amplifies the value of that information.

Existing Intellectual Framework

The information required to launch a new product is assembled within a context established by a specific history of inventions. In addition, each new product requires some level of original invention of its own. As with the parts of the Stanford clock, the value of the information assembled and created by a new product development effort depends upon this existing intellectual framework.

The development of a new luxury car, for example, draws upon the prior invention of engines, transmissions, suspension systems and safety features, as well as air conditioners, mobile stereos and power options. It has value only because an infrastructure of highways exists along with all the associated support technologies. Furthermore, the market success depends, in part, on a particular value set being important to prospective customers. If these things are not in place, the knowledge required to build even the best new car is valueless.

To illustrate the point, imagine members of an engineering team developing the Saturn in the year 1875. A dealership opened on the streets of Dodge City might have created quite a stir, but it would most likely not have resulted in financial success. Even though they may have created the exact product that was actually introduced in 1990, it would have had little or no value.

Development Processes

A major fraction of the effort in a development project is invested in largely routine work that increases the functions of the new product information set in various ways. This work includes processes such as designing a printed circuit board, encoding a software module, or designing a simple mechanical part. In these design examples, functional specifications are transformed into detailed design information. The design processes used are generally familiar to engineers, and the outcome of each is assured. This class of activities is analogous to the manufacturing processes used to increase the function of each part in the Stanford clock.

Congruence of Place, Time and Need

To have value in the product development process, the right information must be delivered, in usable form, where it is needed in time to be used in the product solution. This accessibility of information applies to both outside knowledge coming into the development process and to information flowing through the process itself. Learning of a particular design technique or an applicable technical standard, for example, can greatly influence the outcome of a development project and thus can have great value. If the knowledge is obtained after design freeze, however, it has no value at all to the current effort. To have value, product information generated at one point in the development process must be communicated in an accurate and timely fashion to other places in the process where it is needed. Miscommunication is a common source of waste in product innovation processes and represents a major opportunity for improvement.

Usability

The value of information is dependent upon its form when received and how readily it can be applied to the problem at hand. Engineers often get information in unusable form. Sometimes its sheer volume and diffuse nature cripples their ability to find and apply the needed part. Key ideas scattered widely throughout a pile of textbooks, periodicals, and other reading material may go unnoticed even though they are right there in the office. Gathered in a well written technical paper, however, those same thoughts might have an immense impact.

Occasionally people stop by with bright ideas on how to solve a problem or make the product better. Though meant well, these ideas are often unusable. Time and effort is required to verify their merit and develop them into real solutions. The return on this investment is generally low, so people often discard these intellectual offerings.

Sometimes information comes as vast bodies of knowledge in unrelated disciplines. It is unusable, however, because the necessary intellectual bridges to a current problem are not in place. For example, the premise of this book is that well understood manufacturing practices can apply to the product innovation process. The intent of this writing is to provide insightful connections—the intellectual bridges—that extend the usefulness of these principles into the product development domain.

INFORMATION THEORY CONSIDERATIONS

In 1923, Hartley described the mathematical principles underlying the transmission of information.[1] This work has provided an essential foundation upon which modern communication services rest. Things people take for granted today—telephone service, television, high quality pictures from Jupiter—are possible, in part, because of Hartley's early contributions.

In his work, Hartley described the flow of information through a communication channel with the following equation:

$$I = R \, \mathrm{Log}_2(1/p)$$

Where I is the information flow in bits per second
R is the number of events on the channel per second, and
p is the probability of the outcome from each event.

This equation provides a cardinal insight into the issue of adding value to information. First, it states the obvious—that information flow is greater when R, the rate of information-carrying events, is increased. Most interestingly, though, it says that the amount of information carried by each event, $\mathrm{Log}_2(1/p)$, is related to the probability of the outcome from each event—the lower the probability of a particular outcome, the higher its information value. Hartley further derived the result that, for any number of possible outcomes, average information flow is maximized when the probability of each outcome is made equal to that of the rest.

To illustrate with numerical examples, if a particular event yields an outcome that has a probability of one-eighth, the information value of that result is $\mathrm{Log}_2(8)$ which equals 3.0 bits of information. If the probability is one-fourth, the information value is 2.0 bits of information. An outcome that has a probability of one-half—a fifty-fifty chance of happening—yields one bit of information. Finally, a result that is certain to occur—that is, its probability is 1.0—has an information value of $\mathrm{Log}_2(1)$, which equals zero. In other words, an outcome that is totally predictable has no information value at all! With regard to optimizing probability distribution, if events on a given channel can each have five possible outcomes, then the average information flow is maximized when each outcome has a probability of 0.2.

In the product development process, experimentation is often used to gain or add value to information. Every "breadboard" or prototype is a form of experimentation in which ideas are implemented and tested. The results of these tests yield new information—how well the current design works and which direction to go in making it better. These activities take time and effort and thus are costly. The return on that investment is the information gained. The implication of Hartley's equation is that development energy should be focused on those activities that have an inherent uncertainty. In fact, those elements of a particular development activity that are the least certain will yield the most information. Conversely, investing energy in prototyping parts of the product that have been done many times before on other projects yields little information. Effort spent here has very little return in new information and yet often occupies a disproportionate amount of our attention. Hartley's theory can teach a great deal about improving the information yield of the product development process.

THE KING'S CHALLENGE

A not so practical exercise will illustrate Hartley's information theory and how it can be applied. This exercise takes the form of the proverbial quest to become royalty through marriage. It seems that there was once a king who wished to find a suitable mate for his son and devised a test for prospective young female suitors. His challenge was this:

There are ten coins on the table, identical in appearance. Nine of them are pure gold and one is false, differing from the others by weight. Also on the table is a simple balance scale. Each young lady was challenged to devise a scheme whereby the false coin could be identified without fail through only three experiments with the scale. If successful she would win both the king's son and half of the kingdom. If she failed, she would lose her head in a most expeditious manner. Presumably, the king was interested in culling losers from the gene pool.

Knowledge of Hartley's information theory would have made rich any young woman with a little mathematical skill. Its principles enable the straightforward design of a solution to the king's challenge.

To begin, the amount of information that is needed can be computed. The probability of any given coin being false is 0.1, so the amount of information needed to point with certainty to the correct coin is $Log_2(10)$ or 3.32 bits. The balance scale has three possible outcomes; tilt clockwise, tilt counterclockwise or balance. Experiments need to be designed so that the probabilities of these outcomes are as near to equal as possible to gain the most information from each experiment. If the process is done perfectly, each experiment will yield $Log_2(3)$ or about 1.58 bits of information. Three experiments will then yield 4.75 bits which is more than enough.

In the first experiment, the probabilities can be most closely equalized by dividing the coins into three groups: two groups of three coins each and one group of four coins. The probability that the false coin is in a particular group is 0.3, 0.3 and 0.4, respectively. The obvious first experiment is to place the two groups of three coins on the balance scale. The probabilities that it will tilt clockwise or counterclockwise are both 0.3, and there is a 0.4 probability that the scale will balance. If the scale tilts, the experiment yields 1.74 bits of information and if it balances only 1.32 bits, but still enough to go more than one-third of the way to the goal.

The information from each possible outcome in the first experiment is as follows:

Tilt clockwise—The false coin is heavy and in the group on the right or it is light and in the group on the left. The coins in the group of four are all genuine. (1.74 bits of information)

Tilt counterclockwise—The false coin is light and in the group on the right or it is heavy and in the group on the left. The coins in the group of four are all genuine. (1.74 bits of information)

Balance—The six coins on the scale are all genuine. The false coin is in the group of four. (1.32 bits of information)

And so it goes. Subsequent experiments are designed in a similar fashion, progressively focusing in on the false coin. Coins that are known to be genuine are used as standards in subsequent experiments. Rather than belabor it here, a complete solution to this puzzle is given in Appendix A. In the end, the false coin not only is identified with certainty, but whether it is heavy or light also is determined. This is an additional 1.0 bits of information beyond the demands of the original problem. This information, added to the 3.32 bits required to identify the false coin, makes a total of 4.32 bits of information produced by the three experiments. This result is not far from the theoretical maximum that was computed at about 4.75 bits. The difference is related to the inability to design experiments that have probabilities exactly equal for all outcomes, given the constraints of the problem.

Going through the three tests, the exact information gained can be calculated from the probability of the actual outcome of each experiment. As these amounts are totaled through various possible sequences of outcomes, the total information gained inevitably comes to 4.32 bits. Sequences can be found that do not reveal whether the false coin is heavy or light but that do identify the false coin. Here the total information gained will add up to 3.32 bits, regardless of the path taken through the experiments. Finally, poorly designed experiments will fail to disclose the false coin. They may perhaps narrow it down to a choice between two. The information total for such sequences will be 2.32 bits, one bit short of success.

A Note from Decision Theory

Yet another perspective can be found on the value of information by turning to formal decision theory. An equation from this discipline describes the expected value of an outcome as follows:

$$E = V p$$

Where E is the expected value in dollars
V is the Value in dollars if the outcome is successful
p is the probability that the outcome will be successful

In product development activities, various features of a product contribute specific value to its eventual market worth. The feasibility of each of these features varies depending upon the demands that it places upon existing technology, manufacturing processes, and market acceptance. If the probability of a successful outcome can be estimated, along with the value of each feature, then conceivably the expected value of each critical aspect of a product can be calculated. Progress in the development process can then be measured in terms of the increase over time in expected value.

As engineers strive to prove the feasibility of a specified product feature, they work to raise the probability that the feature can be implemented successfully. As this probability increases, the expected value associated with it goes up as well. The value of information that engineers compile and create as they successfully raise probabilities toward 1.0 is thus equal to this increase in the expected value.

Occasionally, engineers prove that a product is not feasible; that is, through their work they prove that the probability of a successful outcome is zero. This information has value as well, for, without it, the project will continue for a long time before it is inevitably canceled. The value of this kind of information is equivalent to the money saved through early cancellation.

Implications for Project Risk

The King's Challenge provides an analogy to development projects that has interesting implications into the nature of project risk. Before one begins, the outcome of the King's Challenge is uncertain; that is, no one knows which coin is false. However, if the underlying principles of information theory and the paradigm for designing experiments are understood, there is no risk at all associated with this activity. A set of experiments always can be devised that will reveal the false coin. The probability of designing experiments that successfully yield the necessary information is 1.0. So risk is not related directly to the uncertainty of the project outcome. Instead, it depends primarily on the uncertainties of discovering pertinent underlying principles and of finding a suitable design paradigm for experiments. More difficult projects with greater uncertainty of outcome may still be without risk. They simply may require more effort invested in a longer sequence of experiments and other value-adding activities.

The product development laboratory that has a great track record for project success is operating in this mode. If a lab has a history such that nineteen out the last twenty projects have reached the market successfully, then the risk associated with projects there is very low—on the order of a 5 percent chance of failure. This success rate implies that the underlying principles and paradigms that work in that particular business are well understood and quite successful. Even so, engineers and project managers generally will feel that what they do is inherently risky and unpredictable because the exact outcome of each project is quite uncertain in the beginning.

Real project risk is more nearly related to how much change a project must make in the existing intellectual framework to make the new product successful. Having information theory in hand, the King's Challenge requires no further change to the underlying intellectual framework and therefore implies no risk. However, if someone first had to invent Hartley's principles of information flow, undertaking the King's Challenge would be risky indeed. In this case, a major change in the underlying intellectual framework would be required to succeed.

A project may require movement in the underlying intellectual framework along several dimensions, including technology, manufacturing and marketing. If project teams intend to use a new, untried technology in the product, a change is implied in that dimension, usually referred to as invention. If they will apply new manufacturing processes for the first time or drive manufacturing cost to a new low, change in the existing intellectual framework for manufacturing is required. Finally, if the new product will pioneer new ground in the marketplace, perhaps provide a new function that no one has seen before, then change is necessary along the marketing dimension.

The degree of project risk is determined largely by how much intellectual framework movement is required in any single dimension and by whether or not a given project affects more than one dimension at a time. An enhancement effort that updates an existing product with current technology carries little risk. Updating a personal computer product with the latest memory and microprocessor chips is an example. This class of project still fits the information assembly line analogy. Information flows into the process, and value is added to the information in a variety of ways until documentation that describes how to manufacture, use, and support the new product is finally delivered. However, work that fundamentally changes the underlying intellectual framework has not been required and so little risk is centered in this project.

In contrast, development of a fundamentally new platform for that same business carries a great deal more risk. This project will have elements that create significant change in the underlying technological framework: the design of a new microprocessor chip perhaps that runs faster and has twice the density of existing devices. Development teams must create new semiconductor processes, prove a new circuit architecture, and understand new timing relationships. As the teams master the challenges, they raise the understanding of these technologies to new levels. They permanently advance the underlying intellectual framework. The project still may include the more routine elements described in the previous paragraph, but the risk is centered around the steps that require invention.

The success of a project that takes on two dimensions of the underlying intellectual framework is significantly more uncertain. Suppose the project team decides not only to create the new platform described above but to bring it to market at half the price of existing, lower performance products. To shift two dimensions may require redesigning all electronic assemblies to use surface mount printed circuit technology for the first time, changing the mechanical housing to a snap-together configuration, and developing the use of robotic assembly techniques. All of these additional steps require significant movement in the existing manufacturing intellectual framework.

These modifications add risk of their own, and the possibility exists for interactions between the technology changes and the new manufacturing methods that create additional risk. For example, the shift to snap-together construction may cause parts to be made out of plastic that were once made of metal. The higher clock rate in the microprocessor raises the frequency of electronic interference generated by the circuitry. The lack of metal in the housing reduces the level of electromagnetic shielding and allows an unacceptable level of interference to leak from the product. These kinds of problems often are discovered late in the project and must be solved before the product can be sold legally in many countries.

Occasionally, a project will attempt to shift the underlying intellectual framework in all three dimensions at once: a new technology at a challenging price point into a fundamentally new market niche. Start-up companies often find themselves in this situation. When they succeed the payoffs can be immense. The risk in such ventures, however, is extreme, as is reflected in a very high project failure rate.

Project managers can control risk by pursuing a product line strategy that confines any single project to only one risk dimension. For example, recall how companies addressed a new market opportunity, electronic games, a few years ago. Early products were constrained to existing chips, display technologies and manufacturing methods. The risk was confined to understanding the nature of customer demand and defining the right products. Elimination of technology and manufacturing risks allowed quick development of new products and fast turnaround of enhancements. Later, when the market was better understood, new computing and user interface technologies were brought in to enhance product performance. Finally, new display technologies and manufacturing methods were introduced to lower costs and enable miniaturized construction of hand-held products.

STRATEGIES FOR IMPROVING INFORMATION PROCESSES

Select the Best Available Opportunity and Align the Project Well

As indicated earlier, an opportunity's magnitude multiplies the value of information created in the subsequent project. Effort spent identifying and clarifying the opportunity before project activity begins in earnest is thus well rewarded when the product ships. Projects are easy to align with opportunities. The converse is not true. If an opportunity is well understood, the project can be perfectly customized to it in the very beginning. But if the project proceeds without a clear opportunity in mind, the marketplace will not adjust to the resulting new product well at all. As will be discussed in the next chapter, opportunities are not cast in concrete and change over time. Faster projects are thus easier to keep aligned and more likely to succeed.

Provide Visionary Leadership

Vision is often essential in revealing the value in information. A visionary leader helps the project team see the value that their work has to the customer. Most engineers are altruists at heart and are highly motivated to help others with their special skills. A visionary leader clarifies the connections among their skill sets, the information to which they have access, and the needs of the customer. Once these connections are clear, engineers generally will work hard to turn vision into reality.

Work on the Least Probable, Most Valuable Tasks First

As indicated earlier, the least probable outcomes yield the most information. Expected value increases fastest when high value outcomes that have low probability of success are proven feasible. On the other hand, if they prove unfeasible, contingency plans can be initiated early. If no such alternatives can be found, then the project can be canceled at the earliest possible time. Cancelling a project early is sometimes the most valuable outcome possible. Conversely, when a questionable task is left until late in the project and then found to be impossible, contingency plans are more constrained and the cost of project cancellation is much higher.

Strive for Effective Design of Experiments

Most projects include some amount of experimentation to resolve unknowns and to determine the best approach to a design. Experiments should be viewed as learning experiences and optimized to yield the most knowledge. Experiments that have multiple outcomes that are equally probable yield the most information on average. In contrast, tests that are almost certain to produce the expected outcome create the illusion of progress but yield less information. A balance is needed between these trivial experiments and those that are so improbable that they are unlikely to succeed.

Find Ways to Increase the Rate of Doing Experiments

Information flow depends both on the amount of information yielded from each event and on the rate at which events occur. Often, teams can perform experiments more frequently if they eliminate unnecessary steps or frills in implementation. This is an opportunity for creativity.

For example, early in the development of a large drafting plotter, the development team members needed a competitive technical approach. HP labs in Palo Alto had created an 8-inch wide breadboard mechanism that moved paper back and forth underneath the pen. It worked extremely well but, after seeing the mechanism perform, the drafting plotter team leader felt that the basic approach was inadequate to handle D-size sheets of paper. The HP labs project leader ordered the little mechanism to be sawed in two on a band saw. The team quickly welded extensions in place that reconnected the two halves. In less than a week, a 24-inch wide mechanism was operating that proved the feasibility of the concept for drafting plotters. A more conventional approach to the experiment would have involved designing new parts for the wider machine, documenting those designs, fabricating the parts and then assembling the new mechanisms. These steps would have taken months.

Seek and Use All Information Available from Each Experiment

Each experimental outcome yields information in a variety of forms. Careful thought is needed to recognize the full implications of a given outcome. In the King's Challenge, for example, a common mistake is failing to recognize all of the information implied by a tilt of the scale. Often opportunities to learn from an experiment are left on the table, unused. Breadboard mechanisms, for example, are used to prove concepts but are rarely subjected to environmental tests. Without careful, worst-case design, breadboards inevitably fail such tests. Nonetheless, the manner in which failure occurs and the actual limits at which it happens often can reveal sensitivities that better prepare designers for the next step. The extra effort to extract this information is often small and well worth the investment.

Once the design has reached prototype stage, strife testing can provide another way to squeeze information value from the investment, though often by transforming the machine into scrap. This form of testing involves working the machine hard while subjecting it to increasingly difficult environments. Engineers push working conditions incrementally toward extremes, often well beyond the specified limits, until the machine fails. When and how failure occurs provides insight into ways that the design can be improved.

Engineers do strife testing at extreme temperatures, extreme levels of humidity or while zapping the prototype with increasing levels of static electricity. Depending on the nature of the product, testing at extreme levels of shock and vibration also can be revealing.

Some people, though, often resist this form of exploration. Apparently, once they get a prototype working, they want to keep it intact at all costs. An HP colleague Al Kendig said, "Finding a bug in your design is cause for joy and celebration! It reveals an opportunity to learn and move your design forward. And it means that you won't have to fight that particular bug after you release the product to the customer." Strife testing can be a very useful way of accelerating the process of learning.

Minimize the Time Spent on Non-value-adding Activities

Organizational velocity is defined as the ratio of time spent on activities that add value to the total time spent. This metric works as well for development projects. Whenever engineers are distracted—spending time at activities that do not use their skill to add value to information—the forward velocity of the project decreases. Non-project related interruptions are a key problem. Tom DeMarco, a management consultant, claims that each interruption costs at least twenty minutes of lost concentration.[2] Three telephone calls an hour, for example, will just about negate any useful intellectual progress on a problem that requires focused thought. Managers should watch for and eliminate distractions that affect their engineers.

Provide Information Resources for Engineers

Every time engineers encounter a lack of information, they must stop and search for what they need. While locating and delivering information to the place it is needed adds value, using engineers for this task is generally not very efficient. A professional librarian is usually more effective and is an important resource for a development team. Having materials engineering specialists available to find needed parts and materials also speeds the process. A support organization should anticipate the information needs of development teams and routinely make available information on a variety of topics, such as customer needs, the competition, design rules for manufacturability, applicable standards, and preferred parts, to name a few. A common failure mechanism in the development process occurs when engineers proceed without needed information and, as a result, create flaws in the product design. They might ignore an important standard or overlook the special language needs of a critical foreign market. These flaws often go undetected until the product is in the hands of the customer.

Correcting problems at that point is immensely expensive in both dollars and lost customer loyalty.

Design and Manage Project Information Flow

Companies should regard the information created by the engineering team as extremely valuable intellectual property, the crown jewels. As such they should handle it in a secure and professional manner. They need to put a product data management system in place that regulates the flow of and access to this information. Configuration management procedures should ensure that only authorized individuals are able to alter a design. This system should keep track of the current version of each element in the product design as well as any earlier versions that must be supported. In the development of a complex product, opportunities for information-handling errors abound. For example, design effort might be invested in the wrong version of a part and need to be discarded, or an improved version of a design might be discarded by mistake and need to be recreated from scratch. If a design is not documented properly, engineers who must later fix a bug cannot understand what was done by the original designer. The information that describes a new product often costs tens of millions of dollars to create and may be worth hundreds of millions in potential profits. It is the result of dozens of engineering years of career effort by good people. This information should be handled with great respect.

NOTES

[1] Hartley, R. V. L. 1928. Transmission of Information. *Bell Systems Technical Journal:* 535.

[2] DeMarco, Tom and Tim Lister. 1986. Notes from "Controlling Software Projects" from a seminar by the same name.

CHAPTER 7

Lessons from the Manufacturing Domain

Again we start walking. But this time, Herbie can really move. Relieved of most of the weight in his pack, it's as if he's walking on air. We're flying now, doing twice the speed as a troop that we did before. And we still stay together. Inventory is down. Throughput is up.

GOLDRATT AND COX, *THE GOAL*

THE TIME-BASED FACTORY—TIME-BASED PRODUCT DEVELOPMENT

In the July–August 1988 edition of the *Harvard Business Review*, George Stalk Jr. introduced the concept of the time-based factory to the manufacturing community at large. In his article, "Time—The

Next Source of Competitive Advantage," Stalk alerted the rest of the world to the fact that the Japanese had moved on from the focused factory concept and were now pursuing something called the time-based factory. The idea behind the time-based factory is that, since time is of value, factories that can respond more quickly to customers' orders bring more value to the customer and are therefore more competitive.

By way of example, Stalk describes Toyota's approach to delivery. Under the usual circumstances, when buyers walk into a car dealer's showroom and order a custom-configured car—that is, with their specific colors and options—it ordinarily takes eight weeks to deliver that car. Toyota decided this delay was unacceptable and began to attack time wherever it was being wasted. Working from this new perspective, Toyota managers found problems with their order processing cycle, delivery cycle, and factory response. They systematically began to remove time out of all those elements of the cycle. The net result is that customers now can walk into a Toyota showroom in Tokyo on a Monday, order their custom-specified cars, and get delivery on the following Monday.

Toyota's accomplishment has important ramifications for the product development domain. Certainly, customers for the output of the product development function are no more interested in waiting long periods for their needs to be met than customers of a factory. So the question becomes, is there something about the time-based factory that can be mapped through the metaphor back into the product development domain that will enable that function to become a time-based activity? This improvement is the ultimate vision. However, before a time-based product development activity is feasible, a considerable level of development process maturity will have to be attained, as discussed in Chapter 5.

It's useful, therefore, to understand how the time-based factory comes into being. Planning is crucial. Someone has to think through the full cycle, including supply lines and response time from suppliers, work-in-process flow and the elimination of bottlenecks. The same is true in the product development domain. In order to effect a quick-response development cycle, the environment has to be built and the resources must be waiting in the wings to attack time problems whenever they occur.

Actions that Reduce Manufacturing Cycle Time
• Improve process quality (minimize rework)
• Implement concurrent processes
• Add value as rapidly as possible at each process step
• Improve quality and timeliness of incoming materials
• Streamline the flow of materials
• Eliminate work that adds no value
• Minimize change-over time
• Eliminate bottlenecks

TABLE 7-1.

TRANSFORMING MANUFACTURING CYCLE TIME REDUCTION METHODS

The following discussion considers how companies accomplish a time-based factory and then maps these ideas through the metaphor to see how they apply to product development. Table 7-1 lists a number of well proven approaches to removing time from the manufacturing process. Each is examined below through the lens of the metaphor.

Improve Process Quality. An obvious place to start in accelerating factory assembly line performance is to make individual process steps as effective as possible. A good measure of process effectiveness is the amount of rework necessary to make the output acceptable for downstream activities. Output from a high-quality process rarely requires rework and needs little or no testing. Time and energy spent on rework add no new value and simply create waste that never can be recovered. These concepts are true for product development processes as well.

Take the design of an electronic assembly as an example. The input to the process is information describing design requirements; the output is a schematic diagram of the resulting circuitry. Other information used in the design might include knowledge of preferred parts, cost requirements, manufacturability constraints, and information about pertinent performance standards. Engineers executing this process use a variety of methods and tools depending on the task at hand.

Their intent is to create a design that satisfies all requirements and constraints, and then to deliver that information in usable form to the printed circuit design process. Process quality is indicated by the number of times the engineers are called upon to eliminate design defects that have been found in subsequent process steps.

Process improvement can be achieved through well known quality methods, including "root cause analysis" and a "plan-do-check-act cycle." In other words, why did the design go wrong, and what might prevent the same problem from happening again? To make these methods work, however, engineers have to acknowledge that what they do is a process, and they must be willing to change their approach in an effort to improve the probability of successful outcomes. Typically, the project development team discovers potential process improvements too late in the project cycle to apply them on that project. To have value then, other engineers on subsequent projects must accept and use these improvements.. This transfer of learning between engineers working on different projects puts a strain on the individualistic culture that pervades most U.S. engineering communities and perhaps explains why process improvement in product development rarely occurs spontaneously.

Implement concurrent processes. Modern factory assembly lines consist of multiple parallel activities that feed into a final assembly stage. Work is broken into the largest number of concurrent activities deemed practical to get it accomplished quickly. In the auto factory, for example, engines, transmissions, chassis and bodies all are built at the same time for a given customer order, and all come together at the final assembly line. In the product development domain, similar concurrencies are necessary. Everyone works in parallel: One group designs the card cage, another designs the package, and the software engineers design the firmware. At the same time marketing people work the product plan, and manufacturing engineers design production tooling and processes. The idea is to get as many people as possible employed effectively in bringing the product to market in the shortest possible time.

Concurrent product development includes some hazards, particularly the chance of creating scrap. Product development activities contain a greater degree of uncertainty than do auto assembly lines. No team has full information about the outcome of development efforts in other areas, so there is some possibility that various elements which are designed concurrently will not function together.

Consider once again the development of a printer. One team works on the ink cartridge, while another designs the mainframe. Still another team works on the formatter board to convert the incoming data and format it into the electronics that drive the print-head. Another team develops the power supply, and yet another creates the documentation. All of this activity takes place concurrently. Suddenly a snag appears. The print-head people realize that their approach may have a flaw, and they have to change the form factor. Or they discover a functional problem: the print-head must go through a start-up procedure each time the printer is turned on in order to prime the ink cartridge, and a "service station" is required. The mainframe team, however, has already tooled the product, and the service station was never in the plan. So the attempt at concurrency introduced scrap.

Tooling and time have been lost. Not as much time was lost, however, as would have been wasted in a sequential development of each project element. This example brings the discussion back to the original point of this book: innovation cycle time is the most important element in achieving business success. Development cost, on the other hand, is secondary by comparison. Referring to Figure 1-2, for a product with a five-year market window, a 50 percent overrun in development costs is about equivalent to slipping six months in development time in terms of the impact on profits. The cost of scrap in this example represents a small fraction of the total R&D budget, nowhere near 50 percent. Furthermore, the scale tilts even more in favor of risking scrap to save time when market windows are shorter than five years. In general, concurrent engineering of new products is worth the effort and the possible added expense.

Add value as rapidly as possible at each process step. In manufacturing, adding value implies processes and tooling that rapidly perform multiple operations on incoming materials to create major increments in their value at each stage in the assembly line. In product development, this improvement translates into processes and tooling that rapidly create major increases in the value of information.

This area represents an opportunity for real creativity and investment in development processes. Once development teams characterize and understand a particular step in the development cycle, they can find ways to accelerate its contribution to the information set. This work needs to be done individually for each business, so a general discussion here is of only limited value. A few key areas are, however, worthy of mention.

Rapid prototyping is one such area of focus. Prototyping elements of the product design is a common method for eliminating uncertainty and verifying performance. Leading-edge development teams are discovering new ways of prototyping that cut large amounts of time out of the process. For instance, software tools create simulated user interface controls on a computer screen. In mechanical design, a recently developed process called stereo lithography promises to remove large amounts of time in the fabrication of complex prototype parts. Perhaps the fastest form of prototyping is computer simulation. Computer models that behave like real-world mechanisms allow the designer to iterate design parameters in a matter of minutes rather than days or weeks. Commercially available simulation tools and libraries are evolving at a rapid rate and are key resources for accelerating innovation.

One caveat is important. The total time required for a process step is equal to the time required to do the process once times the number of iterations of the process needed to reach the desired result. Occasionally, when the time required for a process is reduced significantly by some new tool or method, engineers respond by adopting an iterative approach to their design. Rather than think through the design in an attempt to get the desired result in the first pass, they take advantage of shorter cycle times and use trial and error instead. The net result can be that expected time savings shrink to zero. Development teams must take care to manage and improve the entire process, not simply focus on individual subprocesses.

Improve quality and timeliness of incoming materials. In manufacturing, incoming materials have been a key area of focus in removing waste and improving quality. In earlier times, companies spent immense amounts of time and energy inspecting incoming materials to ensure quality. Now vendors have improved quality to the point where it is more efficient to detect and replace the extremely rare bad part than to inspect all of the parts before installation.

In product development the need is for delivery of high quality information to the engineering process just when it is required. Engineers need a variety of information throughout the development process; some they can anticipate and some they cannot. Much of the information is generic, such as design-for-manufacturability guidelines or environmental test limits. Other needed information is specific to the current product effort, such as customer requirements or simulation models for specific parts.

The effect of poor quality or missing information on the engineer is the same as that of an inadequate materials supply on the assembly line worker. If materials are missing, the assembly line shuts down until they arrive. If needed information is not available, the product development process shuts down until it is in hand. If poor quality parts are installed in the assembly process, costly rework will be required to eliminate the resulting product defects. If engineers act on poor quality information, similar rework will be required to remove defects from the product design.

Similar improvements are needed in the support systems that deliver information to the product generation process. They may take different forms than those implemented in manufacturing for materials delivery, but the end goals are exactly the same: to deliver the highest possible quality where it is needed, when it is needed and in usable form. Today the information delivery system is often the engineer who needs the information. Once the need is discovered, engineers simply take off their engineering hats, put on librarian hats and go looking for the information. This search may get the job done, but it certainly imposes a time penalty on product delivery. Investing in information delivery resources is a key area of opportunity to streamline product development.

Streamline the flow of materials. Modern assembly lines are characterized by the well-planned flow of materials, and the efficient conveyors and handling equipment that move an assembly from one station to another. Product generation activity should also be characterized by the well-planned and efficient means by which information moves from one stage of development to another. This smooth flow of information implies that the entire enterprise has a consistent, integrated set of information networks and tools. Members of the development team use compatible tools and routinely move information back and forth over local area networks (LANs). Information moves from development to manufacturing in a similar fashion and is received there by compatible manufacturing tools.

Geographically remote members of the team have similar access to information through wide area networks (WANs). In leading-edge installations, engineers can work interactively on joint tasks and instantly see the results of others' work on their own screens. Information added or created by one engineer is instantly available to all. The need for geographic collocation of project teams is being eliminated systematically by modern information tools.

Although many development environments today do not resemble this description too closely, some do. The evolution of information tools has made the scenario described above achievable, and leading-edge companies already are operating this way. Along with other improvements in the overall development process, the information environment can contribute to competitive advantage.

Consider, for example, HP's recent development of an operating system for analytical instruments. Members of the team were located in Waldbronn, West Germany; Avondale, Pennsylvania; and Palo Alto, California. Each group had specific parts of the software to design that were interdependent—changes made in one area affected performance of software in another. The Palo Alto group agreed to maintain the current version of the total operating system and make it remotely accessible to the other sites through a WAN. Each team periodically sent its changes to Palo Alto over the WAN. The next version of the system was then assembled and, at an agreed upon time, made available for remote access. Within minutes both the Avondale and Waldbronn teams had the current system running at their sites, and development continued with all teams once again synchronized. This development effort progressed well and resulted in a successful product.

In contrast, an earlier effort, before the age of WANs, attempted to accomplish the same thing through shipment of magnetic tapes. Shipment delays and time lost passing through customs hampered engineering efforts immensely. Engineers in the three sites only rarely were working with the same version of the operating system. Often as not, recently designed code would prove to be incompatible with operating-system updates that had been two weeks or more in transit. Delays in the shipment of information had similar effects on this development process as had material handling delays that once were imposed on assembly processes.

Eliminate work that adds no value. A generic approach to process improvement that has proven affective in manufacturing is to eliminate work that adds no value for the customer. Each routine task that people perform on the assembly line is scrutinized for its value-added content. Steps that are found wanting are either modified so that they do add value or eliminated. The same principle applies to the information assembly line. Examine the way engineers spend their time and assess whether or not they are adding value to the product information set with each task. Excessive meeting time, for example, is a key candidate for examination. Change or eliminate the tasks that fail to add value. Once again, organizational velocity is the ratio of the time spent on value-adding activities to the total time spent on a task. Eliminating non-value-added work increases organizational velocity.

Minimize change-over time. Traditionally one of the biggest impediments to fast, flexible assembly lines has been the time and labor required to change tooling from one product assembly task to another. Breakthroughs in change-over time and effort have since reduced these costs to an insignificant level in many cases. The result is quicker response to customer orders, lower production cost, and a wider variety of product offerings.

Change-over waste occurs in product development as well. Most notably, the time between when a development team finishes the last project and gets fully engaged with the next is a form of change-over waste. If the background work in market and technology discussed in Chapter 5 is in place, the project team can simply accept the highest priority opportunity in the queue and go for it.

Reduction of change-over time in product development has similar payoffs to those in manufacturing: quicker response to customer needs, lower product cost due to more efficient development and a wider variety of product offerings. The approach to minimizing change-over waste in product development is similar to that of manufacturing; think through the change-over process and design the organization to support rapid change-over of development teams. The objective is to maximize the percentage of time that project teams spend adding value for customers.

Eliminate bottlenecks. In manufacturing, bottlenecks are points in the assembly process that restrict the flow of materials. Speeding up operation of the entire factory becomes a task of identifying and resolving bottlenecks.

Bottlenecks in product development are points in the process where information flow is restricted or the rate of adding value to information is not keeping up with demand. Other members of the project are held up in their work because of the lack of information from the problem area. Speeding up the development effort becomes, again, a process of identifying and resolving these bottlenecks.

In general, bottlenecks are resolved by first identifying the root cause of the limited capacity and then investing the resources necessary to bring capacity to required levels. To prevent unnecessary generation of work-in-process inventory in other parts of the operation, work flow is arranged so that the limiting process paces the rest.

Problems that create bottlenecks may be related to process, input or people, and the solutions vary accordingly. Any operation needs some degree of over-capacity in the wings, however, to deal with bottlenecks when they occur. If this extra measure of resources does not exist, if every available person and machine has been fully committed, then resolving a block in the operational flow is not likely. In this event, the entire operation inevitably will grind down to the rate of flow available through the blockage. As the operation slows down, however, resources free up in underutilized areas of the process that might be brought to bear on the problem.

In product development, one source of extra capacity available to resolve bottlenecks might be those engineers working on investigation projects—small, informal activities that are formative in nature and less time-critical than more advanced projects. A well-balanced lab should have about 10 percent of its work force employed in this capacity. Other sources of extra capacity might be outside consultants or engineering groups that can take on some of the project work and thereby free up internal capacity to take on the problem. Again, spending money on outside help to save project time almost always is a good business decision because of the large impact of lost time on profits.

Sometimes a bottleneck occurs when engineers simply do not have the expertise to complete the work assigned to them. An unanticipated technical problem may have stalled their progress. Bringing additional expertise, either internal or external, to bear on the problem is essential. In any event, a bottleneck in product development implies a large cost in wasted time and effort, and timely resolution is worth considerable attention and investment.

TRANSFORMING MANUFACTURING WASTE CONTROL METHODS

As further illustration of how the information assembly line metaphor maps over from the manufacturing environment, the process of eliminating waste in manufacturing can be compared to the process of eliminating waste in product development.

Manufacturing Waste Redefined

- Line length
- Station cycle time
- Floor space occupied
- Change-over time
- Uncommon dies and tooling
- Inventory—raw, in process, finished
- Lot size
- Worker movement

TABLE 7-2.

In manufacturing, there are four important steps:

1. Redefine waste in creative ways.
2. Enable employees to identify waste.
3. Find waste.
4. Attack waste.

Consider the first step, redefine waste. In the manufacturing domain waste is viewed in new ways. Table 7-2 lists some elements of manufacturing waste of current interest. A few of these are highlighted in the discussion below.

Line length. Line length is defined as the number of steps and time required between the beginning and the end of the assembly process. Associated problems in manufacturing include time lag in shipping to order, amount of work-in-process (WIP) inventory sitting on the line at any given time, excess steps resulting in extra opportunity for error, more hand-offs between steps leading to handling inefficiency, and speed of the total line limited to speed of slowest step.

Product development equivalents include the number of steps and amount of time between the beginning of a development project and product release. Problems include time lag in delivering technology to customer need; WIP inventory invested in product development; large number of sequential steps, which increases the probability of product design error, especially when markets are moving as the design is taking place; hand-offs, which require communication lead-

Actions that Enable Employees to Identify Waste
Educate employees in related concepts:
Material flow
Inventory reduction
Quality and cost of quality
Cost of time
Problem-solving tools
Use of managers as trainers and coaches

TABLE 7-3.

ing to opportunities for information errors; and speed of total development process limited by slowest step.

Station cycle time. The time required to complete a specific step in the assembly process defines station cycle time.

The product development equivalent is the time required to complete a specific step in the design process, for example to design a filter, encode a software module, lay out a printed-circuit board or fabricate a prototype.

Inventory. Raw inventory consists of commonly purchased materials, such as sheet metal, pipe and paint. Work-in-Process inventory means materials that have gained added value in the assembly process but are not yet finished goods. Finished inventory includes the finished products that have not yet been received by the customer.

Product development equivalents—Raw inventory includes investments in technologies that have been invented or market research projects that have been completed that are not yet being applied to customer needs. Work-in-process inventory refers to the current accumulated investment in new product development activities that have not yet resulted in finished products. Finished inventory means completed product designs that have not yet resulted in shipped products, such as system components to be bundled with others that are still incomplete.

These are some examples of how the thinking in the first step of eliminating waste in manufacturing might be applied to the product development process. The second step—enabling employees to identify waste—maps through similarly. Table 7-3 lists some steps that

Actions that Find Waste
• Process mapping • Bottleneck identification • Root cause analysis • Time studies • Adherence to processes • Cross-training, job rotation

TABLE 7-4.

enable employees to identify waste. The discussion below highlights some equivalents in product development.

Material flow. In product development, material flow equals information flow. Employees should be trained to view themselves as agents who add value to information and pass it along. The integrity of this information as it passes through the process is vital, like the accuracy and quality of parts in an assembly line.

Inventory reduction. Management should motivate engineers to move information and technology into customers' hands as soon as humanly possible. Solutions that spend years in the lab under development are of no value to anyone until they are released to production. The useful life of the work is a direct function of how quickly it is done.

Quality and cost of quality. This category includes not only reliability of the finished product but also the elegance of the solution as judged by the customer. In product development, this result depends on the quality of the information used in developing new products and on the quality of the processes used in adding value to the information as it flows through the development process and affects the final design. An early error can be enormously expensive if discovered by the customer.

There are several additional factors in defining and identifying waste in a manufacturing process that have similar equivalencies in product development. The same is true for the last two steps, finding and attacking waste. Tables 7-4 and 7-5 list elements of the last two steps that are viewed as best practices in the manufacturing domain. By now readers can make the extension easily on their own.

Actions that Attack Waste
• Focus on quality, then inventory
• Implement training in quality methods
• Use small lot production, JIT
• Install ergonomic workstations
• Implement quick changeovers
• Shorten production lines
• De-emphasize hard automation

TABLE 7-5.

Once an individual has been trained to identify waste in these terms, opportunities to improve the information assembly line in direct, immediate ways usually become obvious. Not all improvement has to happen in the form of long-term, multiple year programs. Philip R. Thomas, a cycle time consultant, describes these immediate opportunities as "low hanging fruit."[1] The analogy is apt. They are like plums on the low branches of the tree, easy to reach and ripe for the picking. Investing effort in these opportunities yields immediate rewards and often can whet the appetite for going after some of the harder to reach possibilities.

The benefits to this approach in the product development domain are multifaceted. Beyond the obvious improvements in the performance of the activity, plucking the low-hanging fruit reduces the resistance to change inherent in any organization. It immediately validates the value of the changes by providing early benefits in the form of substantial savings.

Low-hanging fruit often abounds in the time dimension of waste. It exists in the form of engineers who are running around unable to do their jobs because they do not have what they need—whether in the form of information or support. For example, if they cannot find the information they need, they go into the research mode and perform what are effectively librarian tasks. Having engineers search for information is waste.

W. Edwards Deming is fond of repeating the aphorism that "leadership means removing the obstacles that keep people from taking pride in their work." Engineers take pride in doing good engineering.

They take no pride in fighting bureaucracy or performing tedious non-engineering tasks. Consider the value of technicians to the engineering effort. Given the opportunity to add staff to the organization, some managers believe an engineer will always make a greater contribution than a technician. The result of that philosophy has engineers performing onerous breadboarding tasks, driving themselves crazy in the process, and perhaps doing a bad job. One technician who could do that kind of work better than they by a factor of three would make the whole organization more efficient, eliminate wasted time, and remove obstacles that keep engineers from taking pride in their work—low-hanging fruit.

Another form of waste in manufacturing that causes a great deal of hand-wringing is work-in-process inventory discussed above. Information systems track it; accountants bludgeon general managers with statistics about it every month, and general managers run down the line chastising people for accumulating it. If WIP inventory in manufacturing includes parts and materials that are sitting around in various stages of assembly, then WIP inventory in product development is information that is sitting around in various stages of completion that has not yet been delivered to the customers who are waiting for it. Every project in an incomplete state in the lab contributes to WIP inventory. A corporate R&D lab of any size probably has $50 million worth of WIP inventory at any time. If projects shrink from, say, 24 months to 18 months, WIP inventory reduces from $50 million to $37.5 million. That reduction may be two to three times the amount of WIP inventory that exists in the entire manufacturing sector of the operation.

For lack of acuity, no one ever pays any attention to this WIP inventory. It has exactly the same accounting impact on the business as WIP inventory sitting in manufacturing. Even though the accountants have not yet learned to recognize it for what it is, it is an immense form of waste. The economic impact of reducing that WIP inventory would by itself justify any attempt at improvement in product development cycles, let alone the increased business success due to improved times to market and the other factors that have been discussed. So the approach described above—redefine waste, enable employees to identify it, find it and attack it—is an extremely powerful tool for improving the chances of business success in the product development domain as well as in the manufacturing arena. The metaphor strikes again.

A useful measure of the effectiveness of quality efforts in manufacturing is "cumulative first time yield." This is defined as the percentage of products that work perfectly immediately after assembly. As total defects go to zero, cumulative first time yield goes to 100 percent.

The product development equivalent is the number of designs that are manufacturable and supportable, and that perfectly satisfy customer needs without rework. The value generally is near zero but not necessarily so. The vision is to improve product development effectiveness so that this metric has a value close to 100 percent.

TRANSFORMING STRATEGIES FOR THE TIME-BASED FACTORY

Returning to Stalk's paper, three strategies are outlined. First, *the move to a time-based factory requires a move from maximum volume production to runs of short time duration that will optimize the flexibility and responsiveness of the factory.* The focused factory concept was based on the idea that orders are accumulated until a really huge volume-production run can be lined up. A huge assembly line is then cranked up to build the products. The motivation behind this approach to manufacturing was the need to amortize the immense cost of the machinery and the set-up time over the maximum number of parts in order to minimize the cost per part.

That philosophy ultimately took factories to the point where, because production lines were so expensive to set up and operate, manufacturers minimized the number of assembly lines and actually reduced the number of models they could offer the customer. They became over-focused. When orders came in for a particular product, they would accumulate until they reached the right size lot, and then the operators would start the line and run all the orders at once. As a customer, if your order was the first one in the queue, you could wait a long time for delivery.

To solve that problem, some number of products were added to each run and placed in inventory. If you were the customer and you got lucky, your order was shipped from inventory. However, there were some serious disadvantages to this approach, not the least of which was the cost of carrying the inventory.

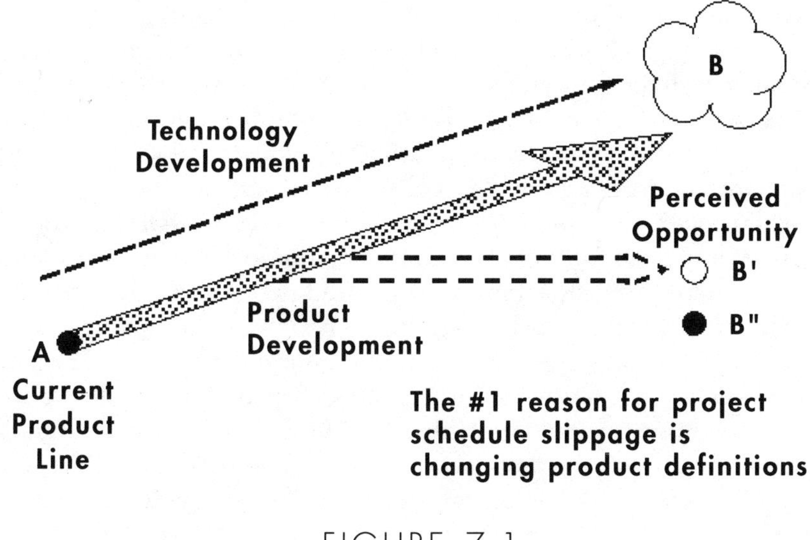

FIGURE 7-1.

The "Long Bomb" Strategy.

In product development, a loose equivalent to the focused factory—the huge production run—is the "long bomb" opportunity. In football, the great dream is to throw that 80-yard pass in the last 30 seconds to win the game. In the factory, the equivalent is accumulating the orders it takes to crank up the machine and run off a million parts. In product development, the long bomb is the development project that attempts to dominate the market in a single stroke.

Referring to Figure 7-1, The "Long Bomb" Approach, the project team begins at Point A with the current product line and a vision of a wonderful, dream-like product at Point B. This dream is way out there, fuzzy, and it is going to require some significant technology development along with the product development necessary to realize it. But when all of the remarkable innovation is complete, this product is really going to make people sit up and take notice. It is going to be wonderful. Of course, it will to take four years to develop, but this investment will be worthwhile. So the long bomb is launched.

Two years down the road, the project team begins to realize that conditions have changed out there in the marketplace. Sometimes they change because of forces in the marketplace itself and sometimes because of technology. Whatever the case, the team now recognizes that the real opportunity is no longer at Point B; it seems to be over there at Point B'. So they make a mid-course correction that causes a lot of engineering to be thrown out. Much of what was designed no longer fits. Product specs have changed; the price, for example, will have to be about a third of what the original projections suggested. And all these factors indicate just the beginning of their changed perspective.

The project team members apply a major reset in the middle of the project and head for Point B'. When they finally get there—four and a half years after project initiation—they find that the real opportunity is at Point B", and they have missed it by a significant distance. This experience is typical for products with long-term development cycles.

As was mentioned earlier, the principal reason for project slippage is changing product definitions. On a four-year, long-bomb project, getting the product out according to the original specifications is virtually impossible. Something simply has to change in that length of time.

This problem is not just hypothetical. A survey several years ago asked HP engineers to identify their biggest obstacles to productivity. The most important problem cited was, "We can't seem to make up our mind, as a company, on product directions and definitions." Engineers were quite aware of the percentage of their time that went into designs that were later scrapped. HP has since invested a great deal in organizational and process changes to address this problem.

Returning to the factory metaphor, the alternative to the focused factory applies time-based approaches. Emphasis shifts toward shorter, incremental projects that deliver essential core contributions to the strategic plan and that provide more flexibility in the flow of product development.

Figure 7-2, A Product Strategy with Hinges, generalizes this approach into a time-based product development strategy. Again, the process begins at Point A with the same vision shimmering on the horizon at Point B. Again there will be some need for concurrent technology and product development, but management insists—with the discipline born of experience—on breaking the product development process into steps, specifically three steps. Three 18 – 24 month projects are planned, beginning with product P1. It will be introduced within 24 months. Because of the short time line, the product definition will not change significantly. Engineers will use available technology, and the design architecture will be modular to enable them to

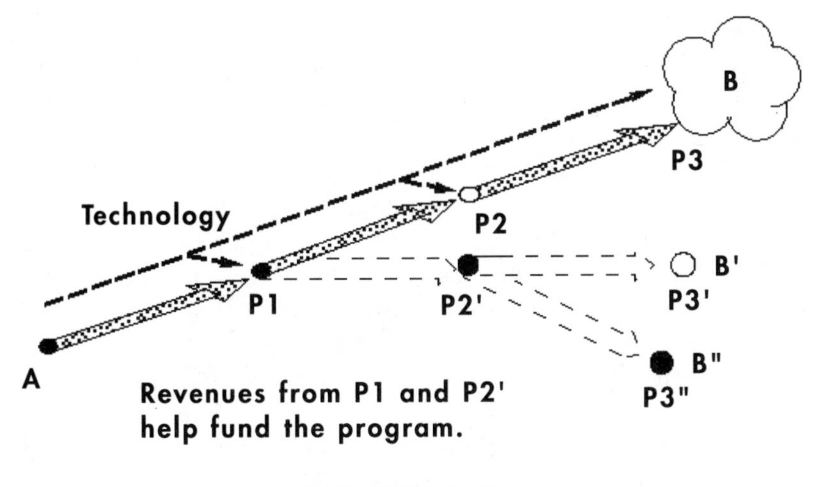

FIGURE 7-2.

A Product Strategy with "Hinges."

reuse major components in anticipated products P2 and P3. With this approach, a product is brought to market more quickly, and progress is made towards the vision at Point B.

With the introduction of product P1, the information-sampling effect described in Chapter 1 takes effect, and the project team receives a rush of information about the marketplace. From that rush of information comes the realization that the opportunity is not really at Point B, but rather at Point B'. So instead of starting on P2 as called for by the original strategic plan, the team modifies the direction and begins developing product P2'. Eighteen months later, product P2' is introduced, and the team receives another rush of information about the marketplace. This new information says clearly that the real opportunity is at Point B". Another 18 months and zap—the project team has arrived at Point B" with product P3".

What has been accomplished with this more measured product development strategy? First, the company has arrived at precisely the market "sweet spot," with the right product at the right price. Second, the company has earned revenues with each new product introduction, and these revenues have helped fund the next development activity. Third, the intermediate products have served to satisfy customers. And finally, the company has established itself as a force in the marketplace and is receiving a continuing flow of information about market directions, in particular about where the next opportunity lies. While it is true that this approach may not suit existing cultural inclinations the way the long bomb approach does, it is a far more powerful strategy.

In *Competing Against Time*, Stalk and Hout describe just such an approach, albeit somewhat accelerated. They relate how the Japanese came to dominate the American residential air-conditioning and heat-pump market. Starting from a position of relative insignificance, between 1979 and 1988 Mitsubishi introduced incremental product innovations into the marketplace *every year*, ultimately arriving at a point where they had gained a seven- to ten-year lead on their nearest American competitors. In the process, they redefined the market by eliminating the need for highly skilled installers, thereby making possible sales, installation, and service by white-goods distributors instead of heating, ventilation, and air-conditioning (HVAC) contractors. This was accomplished without any particular price advantage or startling new technology developments, merely continuing improvements in engineering.

The next strategy in Stalk's *HBR* paper suggests that businesses *move from process technology centers to product-centered organizations.* Looking at this strategy in, say, automotive terms—where a sheet metal facility used to be housed in one factory, a power train facility in another, and an assembly facility in yet another location—there now is simply a Saturn factory.

In the product development domain, the process technology center is equivalent to the traditional functional organization that includes a marketing department, an R&D department, and a manufacturing department. The product development equivalent to the Saturn factory, on the other hand, is a product-centered organization. The project team is composed of all the expertise required to accomplish the project. It must be organized so that communication among all members of this team is absolutely free and easy. Recognition and reward must go to all team members for their excellent collective work in getting the product out. So members of the team forget while they are creating the product that they came from a marketing organization or from a manufacturing organization. Instead each thinks, "I am a member of this product team. My job is to get this product into the marketplace. I report to a manager who is responsible for getting this product to market." Functional boundaries are eliminated that once contributed to the silo effect, the over-the-wall approach.

Stalk's third strategy is *to add value to parts as rapidly as possible, to minimize delays between processes and to strive for continuous flow through the factory.* The factory ought to be viewed as a continuous process. Incoming materials arrive at the shipping dock and are transported instantly to the assembly line. As they arrive, someone turns around and uses them in a product that just happens to be coming by at that moment, and in minutes that product rolls off the end of the line into a shipping container and moves out the door. That case is the ideal. There are no materials accumulating anywhere and no stacks of work-in-process inventory. There is continuous flow, with every step in the assembly process adding enormous value. The completed product is so reliable that it does not even have to be tested, because it always works. Such are the benefits of the time-based factory.

Mapping through the metaphor once again, adding value to information has been discussed already in Chapter 6. Earlier in this chapter, methods were presented to accelerate this value-adding process. Facilitating the flow of information has been discussed already as well. Stalk's third strategy implies that these ideas must be put together in a way that approximates the vision of the time-based factory. Information arrives almost as soon as it's needed. Concurrent development processes simply hum, adding tremendous new value to the information set with powerful tools and processes. Computer simulation is pervasive and almost eliminates the need for physical prototypes that take time to fabricate. Those parts of the product that are intractable for simulation techniques are prototyped in a day or two through leading-edge techniques such as stereo lithography. As the design progresses, a capable configuration management system captures and protects the information. In record time, the design is complete and verified. Documentation that has progressed concurrently throughout the project is complete on the same day. And the project team celebrates another error-free development success.

This smooth process depends on a network of distributed information tools. The local area network is like the conveyors used in manufacturing to move assemblies through the process smoothly. In a similar fashion, the LAN smoothly moves information to the place it is needed. Each engineer has a workstation that can become exactly the tool needed at the time to add some specific value to the contents of the product information set. Appropriate software tools simply are loaded when needed through the LAN from a central library.

The workstation also serves as a window to the world of information, connecting engineers to information wherever it may be found and connecting them to needed expertise anywhere in the world. In a well-designed information environment, engineers access parts information through the LAN while working on their design with CAD tools. Manufacturability information is there too to keep them within the bounds of actual processes available on the manufacturing floor. Information is readily available on design work done by others on the project, and simulation libraries provide standard models for the parts that an engineer might use in a design. Simulation servers are connected to the LAN and stand ready to execute simulation runs for the engineer in the shortest possible time.

Finally, looking just a little way into the future, just-in-time education will be available soon to engineers at their workstations. Occasionally, engineers involved in a project discover a need for knowledge that they do not have, some design technique or interface standard that they have never studied. The lack of this knowledge stands between them and successful completion of their assignment. Soon they will be able to order an educational course on needed subject matter from a central source using their workstations and have it delivered within hours in digital form to a local file. Like a video tape, this course will teach them what they need to know when they have the time to devote to it. At their convenience, they will then use their workstations or other portable information appliances to access this information whenever and wherever they want.

NOTES

[1] Thomas, Philip R.; *Getting Competitive*, McGraw-Hill, 1991.

CHAPTER 8

Management's Role in Innovation

People are inspired to improve **if** *they believe the performance improvement goal is desirable* **and** *realizable, and* **if** *they can see a believable methodology to reach that goal in a reasonable time.*

PHILIP R. THOMAS, *GETTING COMPETITIVE*

Returning to the cash flow wave form introduced in Chapter 1 and repeated here as Figure 8-1, what can management do to improve the shape of this wave form? Given all the principles outlined in the foregoing chapters, the answers by now are apparent. This chapter reviews specific actions that need to occur, explores their impact on the cash flow wave form and outlines management roles and responsibilities that affect the success of the product development process.

MANAGING THE CASH FLOW WAVE FORM

As stated in Chapter 1, the two purposes of business at stake are to satisfy customers and create a good return on investment. Table 8-1

Ways to Improve the Cash Flow Wave Form

- Reduce the time to perceive opportunities (T_p)
- Reduce the time to begin project activity (T_b)
- Decrease the duration of the development time
- Optimize the amplitude of the investment cash flow
- Move release time (T_r) forward in time
- Increase the steepness of the initial profit slope
- Increase the amplitude of the return cash flow
- Move extinction time (T_e) out in time

TABLE 8-1.

lists improvements to the cash flow wave form that will accomplish these objectives. Each item then is addressed in the discussion below.

Reduce the Time to Perceive Opportunities, T_p

First, a clearly defined business strategy—the force field that drives the company forward—is an absolute prerequisite to reduce the time to perceive opportunities. Supporting the strategy are the on-going background processes: technology and market research. A company that does not have, for example, a continuing market research program in place might be tempted to go out and do a spot check of the market before proceeding with a product development program. This approach to sampling the market is fraught with error. Because it is a short-term effort, it can produce wrong answers easily. The most productive approach maintains market research as a continuous program in partnership with technology research efforts. This ongoing contact with the marketplace develops the depth of insight needed to be competitive. The one focus is to queue up and prioritize the opportunities that exist with respect to the business strategy.

Returning to the cash flow wave form, the market research background process accomplishes two objectives: It moves the perception time on the cash flow wave form closer to the opportunity and shortens investment time. With the opportunities already identified, project teams hit the ground running toward the right opportunity and do not have to waste a moment trying to identify an opportunity themselves. Shortening the investment time means faster times to market, which is really the objective of the whole product development process improvement exercise.

Similarly, on-going technology research must be taking place, so that when the marketing people ask, "Wouldn't it be nice if...," the technology people can respond with, "Funny you should mention that. We just happen to have..." That discussion triggers investigation of the next opportunity. Integrating the research and marketing views of the world is an important step in identifying opportunities and requires involvement of individuals who can operate well in both marketing and technology camps.

Reduce the Time to Begin Project Activity, T_b

Once the opportunity queue is established, investment can begin on the highest priority item when the next project team becomes available. Product development will begin sooner if the product development organization has established a pattern of quick, well-staffed projects. If a lab is running six 24-month projects, a team will become free about every four months on average. On the other hand, if projects are all on the order of 48 months in length, an opportunity will have to wait an average of about eight months for action.

Opportunities have finite lifetimes. If they wait too long before an investment begins, they lose their value. In an ideal situation, the background activity that identifies opportunities will work so well that the queue will be long and each opportunity will be truly exciting. The organization will not have the resources to staff projects to address each opportunity and inevitably will have to discard some. Interesting metrics for this part of the operation are (1) the quality of discarded ideas as reflected in the anguish that occurs when they are dropped, and (2) the percentage of great ideas on the top of the priority list that die before they can be staffed or that miss their prime market window because they were staffed too late. An operation is in really good shape when it has so many dynamite opportunities to pursue that it routinely discards great product ideas due to limited resources.

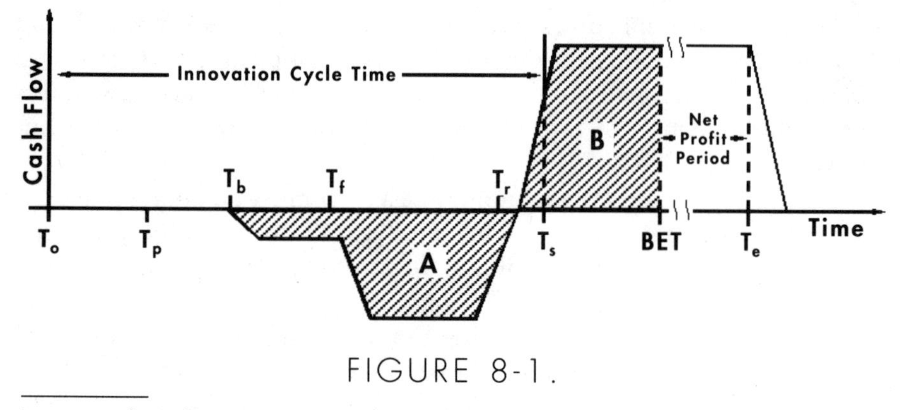

FIGURE 8-1.

Innovation Life Cycle.

Decrease the Duration of the Development Time, $T_r - T_b$

Applying well the principles outlined in the preceding chapters decreases the duration of the development time. Development time is a function of many interrelated factors that include the following:

1. Number of engineers on the project
2. Available engineering expertise
3. Size and nature of the development task
4. Clarity of the opportunity and product definition
5. Level of project risk
6. Development tools available
7. Quality of development processes
8. Information resources available.

If these factors are well managed, development times will be competitive.

Optimize the Amplitude of the Investment Cash Flow

The aim of this item and the one preceding is to optimize the investment phase of the project for business success. Making the most of the investment phase means minimizing the money spent while maximizing the positive cash flow and satisfying customers early—all at the same time. Simply reducing the amplitude of the investment, however, can cause the development time to extend and may not optimize return on investment at all. Each project has an optimum rate of investment that will result in the best business success if the money is well spent. There are no formulas for success here; it is a matter of judgment. Knowing how to invest wisely is, in fact, a key element in the art of management that sometimes comes with experience. More will be said on this topic later.

Move Release Time, T_r, Forward in Time

Release time is really a dependent variable determined by the times to perceive an opportunity, begin the project, and complete development. If those times are improved, as discussed above, release time will indeed move forward in time.

Increase the Steepness of the Initial Profit Slope

Another aspect of the cash flow wave form amenable to being managed is the shape of the return, in particular how quickly it ramps up to full volume. What controls the steepness of the ramp-up to volume production? A variety of factors, the foremost of which is the quality and completeness of the product development effort. The project team has created an information package that describes how to build, use, and support a new product. If a ramp-up to full volume production is required, development of the product has not been finished yet. To say, "Well, we'll ease this product into production," is to say that the development process will be finished on the manufacturing line. The product information set is really at some unknown state here, and the last bit of needed information will be gathered from the assembly line. While this practice is widespread, it reflects the need for improvement of development process quality. The production line is an expensive place to learn, and finishing the development process there also wastes time. Revenues come from volume production; the more time it takes to arrive at volume production the more money is left on the table.

Part of the product development process is preparing the market in advance to buy the product at full volume the day it hits the market. The implication is that the outbound marketing folks have done their job; the dealer distribution channels have been established and are prepared to sell the product, and the customers are ready to buy. The assumption is that the project manager is managing a cross-functional effort that encompasses R&D, manufacturing, marketing at both ends, research, and sales. The rate at which orders go from zero to maximum is controlled largely by how well the outbound marketing job is done. The rate at which production goes from zero to maximum is controlled by how well the transition from development to manufacturing is managed. Both of these ramp-ups are important. A business cannot ship without orders, and even if customers buy millions, the business cannot ship if the product is not ready.

Increase the Amplitude of the Return Cash Flow

An obvious way to increase the return on investment is to amplify the magnitude of the positive cash flow created by the product. This magnitude is also a measure of customer satisfaction in that it directly reflects the value they place on the contribution of a new product. If the contribution is high, customers will support a higher profit margin and buy larger volumes. This aspect of the cash flow wave form is most directly affected by the nature of the opportunity selected, the excellence of the product definition, and the cost and quality of the resulting product. As was noted in Chapter 6, the size of the opportunity multiplies the value of the information associated with a new project. The key to return on investment is to identify excellent opportunities, align the product definition and plans well, and then execute the product implementation in a world-class manner.

The amplitude of the return cash flow is established also by the product marketing effort associated with a project. Identification and education of the sales force and distribution channels that will move the product from factory to customer are critical. Customer support for the product helps sustain a high sales volume. Efficient order fulfillment is important to customer satisfaction and helps keep sales up. And, of course, an efficient manufacturing operation speeds order fulfillment, keeps costs low, and sustains high profit margins. All of these elements of the organization must work together to achieve high amplitude in the return cash flow.

Move Extinction Time, T_e, Out in Time

At the far end of the cash flow wave form, what can be done to affect extinction time? As was mentioned in Chapter 1, extinction time is, to a large degree, a function of product definition. What functions, technologies, manufacturing processes, and market plans will make this product the most competitive answer to the existing opportunity? Beyond the definition, the product implementation must be superb, so that when the product does appear, it leaves the absolute minimum opportunity possible for the competition. Price is competitive, and quality is world-class. Extinction time is thus a reflection of process quality throughout the entire development activity, from definition through design to volume production. As this total process is improved, extinction times on the products that flow through it will move out as far as possible.

MANAGEMENT ROLES AND RESPONSIBILITIES

Project Management

Upper management effectively hands project managers a product development process to manage. It then becomes their responsibility to establish the best implementation of that process. They should consider the tools and processes they have available to bring to bear on the task. In a sense, they need to build an information assembly line that applies the processes already developed by their organization, but they have to customize it to suit the particular needs of their project. So project managers are the assemblers of the assembly line, meta-assemblers.

Assembling the team. Project managers also have primary responsibility for the product definition. It is their job to see that the definition and system design are done well. They should insist on a strong cross-functional team at the earliest possible moment to undertake the definition and the product system design. The importance of this task cannot be overstated, because most of the cost of the product and most of the quality and functions will be determined in the first 15 percent of the product development cycle. Product definition requires the best talent available—comprising R&D, manufacturing and marketing people—working together on the definition and the system design; the architecture; and the partitioning among mechanism, electronics, microprocessor and software. A company runs the risk of flawed results if it does not have the strongest possible team working on product definition.

Unfortunately, assembling the team and harnessing it correctly appear to be the most difficult tasks project managers face. Typically, the first people to become available for the next project are R&D engineers who recently have finished another project. Rather than waiting for help from marketing and manufacturing, on their own they begin to define the product, come up with a system concept and leap to design. Eventually, the project attracts the attention of marketing people. They take a look at R&D's new activity and decide that it makes no sense at all. "You forgot this; you missed this customer completely; there's a whole niche over here; it costs too much, and you're coming out with the wrong product." This interchange results in a big midcourse correction, with significant accompanying waste.

It is up to the project manager to insist on the strongest possible engineering team right at the beginning, in order to do the front-end tasks well. Team members work together to identify applicable technologies, talk with customers and explore manufacturing approaches. They integrate the information they find into a definition of the most competitive product that addresses the opportunity, and they verify it with further customer visits. If the front end work is done right, the project is set up for success with minimum time to market from the very beginning.

Now, how do all the talented people happen to be available at time zero? Resource management. Earlier discussion addressed the need for a judicious amount of over capacity. As in manufacturing, the mind-set has to be broken that insists on utilizing every resource to its maximum capacity 100 percent of the time. Even though it is counterintuitive, booking every resource to the limit usually results in a huge waste of effort. If everyone is always busy on urgent tasks, people are not available when they are needed to resolve a bottleneck, and critical cross-functional talent is not at hand to get the next project started in the right direction. But if managers provide just a little slack—for example, by hiring an outside firm to finish the outbound marketing job or by bringing in people who are presently engaged in some of the background activities—they can assemble the right resources at the front end of the new product development process. Judicious overcapacity may be a hard sell, but it is a major element of any successful program for continuous improvement of the product development process.

Assigning the work. Project managers should assign work on the most critical and uncertain elements of the project first. People like to do what they know they can do, because it gives them the illusion of progress. In addition, management sees early success and is more likely to buy into the project. However, as earlier discussions of information theory have shown, working on the most uncertain aspects first adds maximum value to information in minimum time. Something that is critical to the success of the project should be done first, because if that critical element turns out to be the reason why a project requires an alternative approach or cancellation, project teams need that knowledge as soon as possible. An inevitable cancellation at the earliest time is an extremely productive act.

Prototyping. The objective, as always, is to add maximum value to information at minimum cost at every stage along the line. One of the common failings is associated with the prototyping process. Again, to show progress early, engineers like to make prototypes look as similar to the finished product as possible. A better approach is to use whatever it takes to bring a prototype to the table quickly, because the purpose of the prototype is to learn from it. Usually, no one learns anything by making it pretty, except how much the visual enhancements cost. The prototype becomes scrap as soon as the development team members learn what they need to know from it. There is no value in the prototype except the information it conveys. It falls to the project manager to keep prototype activities as simple and straightforward as possible.

The only exception to that guideline is when development team members need prototypes for customer feedback. They can get some extraordinarily valuable feedback from early users who can provide an unrivaled learning experience. Project managers should think in terms of return on investment of prototypes and information gained per unit of work invested. They should be accountants on that subject and manage the investment of work units for maximum gain of information and the addition of value to the information set. The power of the information assembly line schema lies in its ability to put the project managers' mind-set in the right channel.

Support resources. Support resources could include the entire design environment, from the folks who keep the CAD systems running to the library personnel who find and maintain information critical to the completion of the product.

One of the most important—and perhaps underrated—support resources is the work environment. The project manager must be thinking, "How can I get the most out of this team? What kind of environment do I need to create here?" Some projects, for example, have so many details and so much communication required that the use of a "war room" begins to make sense. Team members set aside one entire work space where they can hang things on the wall that people can update continually. Project managers need to ask themselves if a war room is the right approach or if networked workstations will be sufficient.

We know of one project team whose members decided that the design environment was too full of interruptions. They chose to explore other possibilities and looked for a place that was quiet and free from interruptions to improve their productivity. They discovered that booking a conference room in the building, given the current corporate accounting practices, would cost about $2000 per month. One of the engineers observed that amount was about twice what an apartment rented for in the area. So they rented an apartment, saved $1000 a month and gained the additional advantage of a great deal more room. They set up the apartment with call-out only phones and spent part of each week in the plant attending meetings, handling coordination, and taking phone calls. Although the team members took a lot of ribbing as they moved to this apartment, they had the last laugh. The experience was extremely positive; their productivity was higher than they ever dreamed, and they proved the point that the work environment has a major influence on the productivity of a project team.

Customer links. Absolutely nothing is as valuable to a project manager throughout the whole life of the project as customers with phone numbers. They give development team members a place to go when they have a prototype running and want feedback to verify that they are on track with their product development. As the product metamorphoses, feasibilities and good ideas may turn out to be less so. It is useful for the team to determine that a proposed change is not going to trigger some troublesome reaction in the customer base. Once again, the project manager is the person who establishes and maintains those links.

Waste control. While Chapter 6 covered waste control, this is the place to point out that it is worth project managers' time to figure out what constitutes waste in a particular project; that is, what are they going to define as waste? Waste can be managed creatively. For example, in a software project, engineers can trigger a build of the package, which is a time-consuming exercise. Project managers may want to set up certain specified times for builds that roll in all the most recent modifications. They set up a system in which people learn to work in a pattern, knowing that the next build will be at a particular time. They then prepare for it accordingly. Less time is wasted and efficiency improves.

Portfolio-level Management

The project manager is concerned about a single instance of a product development cycle. Meanwhile, elsewhere in the organization, another project manager is focused on a completely different project, probably in a different time frame than the first. Looking around the operation, there may be four or five project managers, all working on their own instances of the product development cycle. Who is managing the project managers, and putting their projects into play? In some organizations this responsibility falls on the R&D functional manager. This person, however, can readily optimize processes and performance only within the R&D function. There are a number of management responsibilities at the product portfolio level that are extremely important to long-term business success and that affect resources and processes that cross functional boundaries in the organization.

Managing investments. Someone at this level should be coordinating and managing the portfolio of new product development activities. The consideration here is both alignment of the product portfolio with the business strategy and balance for optimum business success. Short-term incremental efforts must be balanced against long-term revolutionary new platforms that will provide the basis for new product families. A balance must be struck between having every available engineer involved in bringing products to market quickly and keeping a small fraction of the work force engaged in investigation of new opportunities. Balance is needed between betting all resources on extremely risky ventures and being too conservative. As in the management of a portfolio of stocks, there should not be too much emphasis on any one area at the undue expense of others.

Project staffing. Part of the task of managing investments is allocation of technical staff to projects in the portfolio. In most cases, the person who manages product portfolio investments will also control assignment of engineers to projects and the movement of engineers between projects. Project managers then will determine the specific work assignment for each engineer assigned to their project.

The number of engineers assigned to a project has a major impact on both time to market and financial success of that effort. The manager in charge at the portfolio level must decide how to spend a fixed budget of human resources most wisely. Maybe the work force should be spread thinly across a large number of projects. On the other hand, perhaps there should be only two projects with dozens of engineers on each. These choices will have a big impact on the business success of the operation. The discussion below explores the relationships that underlie good staffing decisions.

In an ideal world with no secondary considerations, the time required to complete a project equals the total number of engineering months required divided by the number of engineers on the job. The reality, however, is that there is an optimum range of staffing outside of which inefficiency increases and extends the product development cycle time.

This range is reflected in Figure 8-2, which presents data produced by a computer model that simulates typical staffing relationships. The behavior of this model at both high and low staffing levels is adjusted to reflect the author's judgment and experience. The resulting graphs are offered only to illustrate some important relationships and should not be used to predict the performance of actual projects.

In the case of too few team members, people have to concentrate on too many tasks. In addition, as the time to market lengthens, the likelihood of interruptions and definition changes increases. New technology developments intrude that have to be rolled into the design, creating scrap and extra work. The result is that projects involve more work than the predicted number of engineering months, and they take longer than expected to complete.

On the other end, putting too many people on the project can extend its duration as well by imposing an extra communication burden. This effect was evident in the sequential completion of two HP projects. The first was a new drafting plotter mechanism that required 127 new mechanical parts and the efforts of six mechanical engineers over a period of three years. In the follow-up project, a low-cost implementation of the same product, the task was to reduce the number of mechanical parts by two-thirds, thereby reducing the manufacturing costs. That job eventually involved 17 mechanical engineers and took four years. What happened? One engineer on the project said, "Keeping the design straight with so many engineers took too many meetings. We spent most of our time talking to each other and too little time engineering."

Outside of an optimum range of staffing levels, efficiency drops off for both very large and very small numbers of engineers, as reflected in Figure 8-2b. And this fact leads to an important conclusion, illustrated in Figure 8-2c. The return factor has a definite peak and is optimized near the point of minimum time to market.

Although these figures are hypothetical and based on the author's experience, the experience of others also supports the conclusions with ample evidence.[1] In addition, it is possible to support these conclusions by tracing them back through the metaphor to the assembly line. Each assembly line has an optimum staffing level. Too few people with too much to do slow down the line because they are spread too thin. They have to switch tasks, supply themselves with materials, and maintain their equipment. In the simplest example, electricians often say that two people can get a job done much more than twice as fast as one person working alone. And some electrical jobs can't be done by one person.

By the same token, too many people on the assembly line begin to get in each other's way. Too much handing off takes place. Passing materials around the assembly line is exactly analogous to communicating on the information assembly line. Communication waste may well go up geometrically with the number of engineers. So Figure 8-2 describes a common phenomenon with which manufacturing people are familiar but that product development people have not assimilated yet, by and large.

In the end, however, only judgment, developed over long experience, will give managers at the portfolio level the intuitive feel needed to determine the right staffing level for each project. Nevertheless, they should be aware that the over-staffing phenomenon exists and be wary of adding too many people when the schedule starts slipping. More

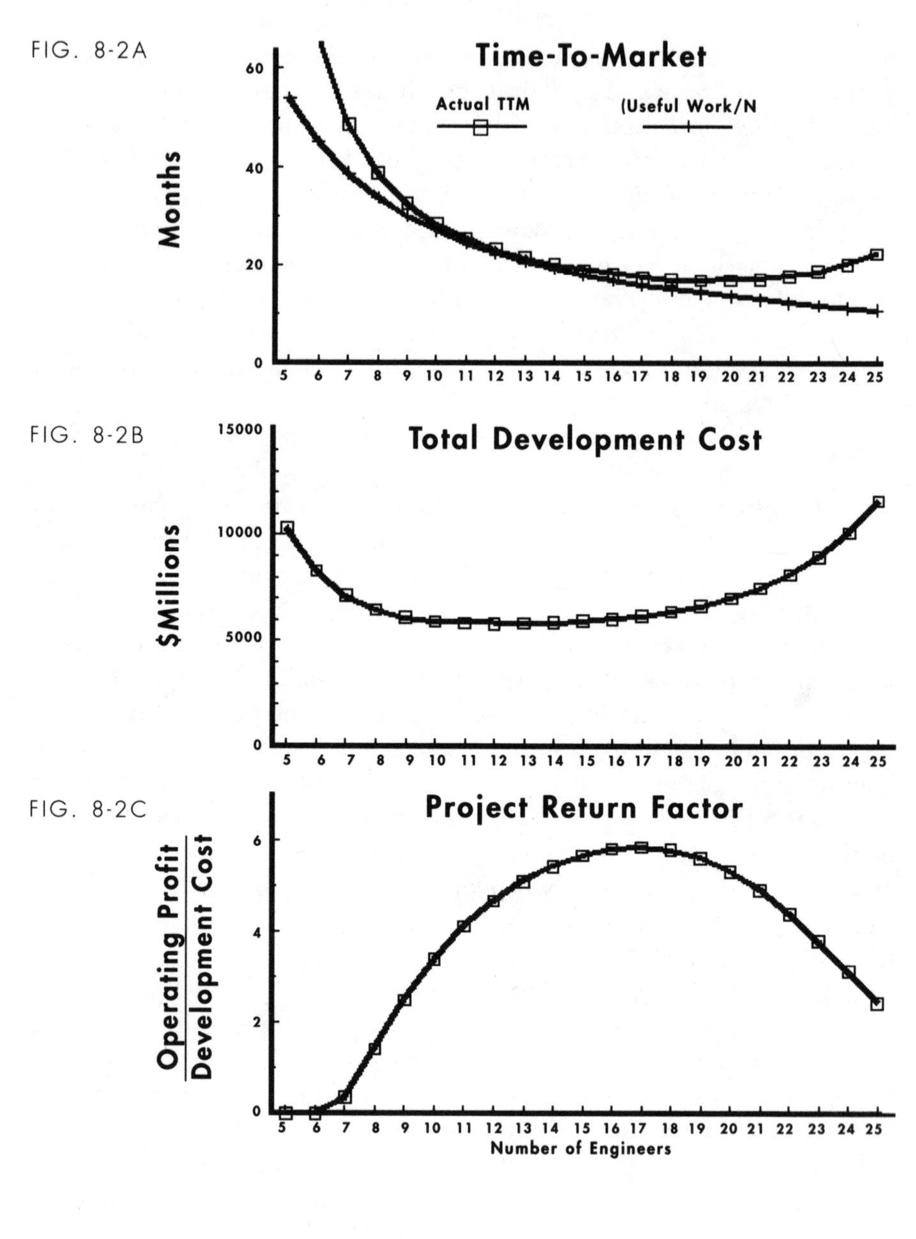

FIGURE 8-2.

Typical Project Staffing Relationship.

people may not solve the problem and, in fact, may compound it. Portfolio managers should keep records of the number of people assigned to each project, its time to market and its return factor, in order to develop the skill required to set optimum staffing levels for each particular business.

Improving development processes. A manager also is needed at the portfolio level who is responsible for the on-going improvement of the cross-functional product development process. This person must have the influence required so team members in each of the functional departments involved with new product development take his or her recommendations on process improvements seriously. This manager must view the process from a sufficiently distant perspective to recognize its flaws and then initiate corrections. Concurrently, this manager must consult with the people involved in the day-to-day operation of the information assembly line to get the same kind of insights that similar consultations with factory line workers reveal in manufacturing.

Information environment. At the portfolio-management level, someone also should be responsible for setting the information environment within which all projects operate. As product development projects proceed, engineering teams require information—about the procurability of parts, for example. The availability of a part that plays a critical role in the design should be determined early on. Otherwise the team risks building a bottleneck into the process. So procurability information has to be available readily.

Another information consideration is design criteria for manufacturability. All projects should have access to design information that ensures that the products will be manufacturable when they hit the production line. This information will be common across all projects for a given business and will reflect the needs of the key manufacturing processes that help make that business competitive. Engineers must design within these constraints in order to create the most competitive products.

Project teams also will need to know in advance what standards apply to a given product. They need to consider and embody standards into the product at the definition stage. If an engineer needs information on some kind of interface standard and has to become a librarian to find it, waste has been built into the product development process.

Engineers also need information on environmental health and safety concerns. For example, engineers may need to verify that ingesting the ink from printer cartridges will not cause undue bodily harm, and they also may need to establish that the residual ink and other materials will not harm the environment after the cartridge has been used. Project teams also must track the environmental impacts implied by the manufacture of components selected for use in the product. For example, there are special taxes on companies that use components requiring CFCs for their manufacture. If the wrong component is specified, a tax might be triggered that becomes an unexpected burden on operations, all because a "dirty" supplier was used.

Different countries have different approaches to environmental considerations, including not only ingredients but disposability. Unless someone is keeping current on the topic, the engineer's job could end up involving a certain amount of political science. The manager responsible for the information environment at the portfolio level must make sure this information is available in a timely way to all projects. Finding out that a product fails to meet Germany's disposability standards after the product is introduced can eliminate a major market and the potential return from it, when the solution could have been incorporated into the design at the outset.

For smaller companies, where these staffing considerations may be prohibitive, some of the functions may reside in one individual, perhaps at the vice-president level. Regardless of where the responsibilities lie, however, the portfolio level functions described here are critical to business success and should be included in every organization chart.

NOTES

[1] See, for example, Fredrick Phillip Brooks, *The Mythical Man-Month*, Addison-Wesley, Redding, MA, 1982.

CHAPTER 9

Summing Up

Whether you think you can do a thing or not,
you are probably right.

HENRY FORD, 1863–1947

What does it take to succeed in a global marketplace? Are all markets inevitably global in nature? We haven't seen much non-U.S. competition in our business. Why should we care about global competition at all? Our business has been successful up to now and, in fact, this has been a pretty good year for us. We've introduced a couple of really great new products that promise to do well. Why should we be concerned at all with radical efforts to improve processes that we are comfortable with and that have served us well for many years?

The questions and attitudes above are common. In some industries they may rightly reflect a steadily growing, successful future. In others, they simply may presage the calm before the storm. In either case, they are based on an adage that passes for folk wisdom in this country. Casey Stengal said, "If it ain't broke, don't fix it!" This book began by describing business conditions in world markets as being similar to a war. In war, if corrective action waits until something breaks, the war will be lost.

The problem is that no one can ever discern for sure between a tranquil moment in a time of continued success and a lull between lethal engagements. The fact is that world markets indeed are becoming increasingly global. In one industry after another, markets around the world are fusing, and the number of key players in each is shrinking. Companies that used to enjoy a comfortable niche in a local market are confronted suddenly with intruders from abroad who do not play by the well accepted local ground rules. (Mitsubishi's incursion into the residential air-conditioning market, as described in Chapter 7 is an example.) Many of these intruders have spent years developing products and services in other areas of the world and only now are turning their attention to opportunities in the United States. While they may be new here, they are anything but start-ups. They know what they are doing, and they know how to enter a new market against local competitors. Their products and prices are good, too good. And, too late, local businesses realize that the deceptively tranquil market that they enjoyed last year was a sham, and they are in for the battle of their lives.

Out of bitter experience, a new wisdom is emerging: If it ain't broke—fix it! The most effective defense against the invasion described above is for local companies simply to be so good at what they do that invasion of their markets looks downright uninviting. This excellence implies that they understand their customers better than anyone and that they are diligent in finding new and more effective ways to serve those customers. In other words, even though their current systems work well, they aggressively invest in making those systems better anyway. In fact, as they grow in their ability to serve customers, they might get to the point where they can do a little invading of their own.

What does it take to be a global competitor—one of the few best sources worldwide for a particular product or service? Once again a metaphor provides answers.

Being a leader in a global business is much like being a winning team in worldwide athletic competition. Intense investment is required in a number of factors that create the winning framework. They include the following:

1. The right equipment
2. Superb talent
3. Great coaching
4. Understanding of fundamentals
5. Hard work, constant practice
6. Continuous improvement
7. Willingness to look at the reality of performance

8. Benchmarking to learn how others do it
9. Understanding the competition
10. Inventing new paradigms of play
11. Loyalty to the team
12. Willingness to withstand extreme discomfort in the effort to reach a goal
13. Eagerness to throw oneself 100 percent into the fray
14. An unshakable winning spirit.

The best teams in the world have mastered all of these factors. Remove one of them, and the description fits the second- or third-ranked team. Remove several to describe the teams in last place and out of contention.

Comparing this list to the current state of new product development practice in the United States provides insight into areas where firms must focus their attention to become more effective competitors. A number of the items on this list are well in hand. U.S. companies, for instance, exhibit genuine leadership in the first two factors. The tools available to Americans truly are leading-edge. U.S. engineering talent is second to none. There are so many examples of U.S. leadership in engineering, from moon jaunts to submicron chip designs, that they defy comment. Performance by U.S. companies on some of the factors listed, however, is sorely in need of attention. This book has explored in depth several of these areas that will be summarized below.

Coaching is one of the more serious problems in new product development. Management's job is to observe engineers working the development process and coach them to better performance. This simply has not been done well, and often little attention is paid to this responsibility. In terms of the athletic metaphor, most businesses have a team manager but no coaching staff. Every business needs coaches in order to survive over the long term. This book has examined ideas that provide the framework for the coaching process: the improvement cycle for product development and techniques for managing improvement programs.

A coach needs to understand the fundamentals of the game. In general, few people understand product development fundamentals well. Although many companies have repeatable development processes, they execute them unconsciously to a large degree. In many businesses these processes have changed little over the past thirty years. Too often businesses do not understand why they do things a certain way. They simply redo the process with which they are familiar because they know it generally works.

Athletes learn by doing—practice, practice, practice! In new product development practice means cranking out new products and executing that cycle as many times as possible. Shorter innovation cycles that are repeated more often accelerate the learning process.

The next three items on the list—continuous improvement, self-analysis, and benchmarking—are simply well-known quality methods applied to the field of athletics. These methods have not been applied to the product development process traditionally, and a key point is that they are as applicable in product development as they have been in manufacturing. The techniques and examples provided on cross-functional product generation processes are essential if companies expect to achieve world-class competitive stature.

Competitive forces in the marketplace drive the improvement cycle described in Chapter 2. Development teams need to understand these forces to determine the issues they must address to stay competitive. As with athletic teams, product development teams need an enemy to defeat. It focuses their direction and puts urgency in their actions. Chapter 8 mentioned that project managers need customers with phone numbers. They need competitors, too, who are just as tangible and just as much a force in their lives.

A final element in this framework for winners is the invention of new paradigms of play. Over time, the teams or athletes who invent new paradigms that enable better performance within the rules of the game gain sizable and immediate advantages over opponents. These fundamental shifts in approach create spurts of rapid progress in performance levels that move the entire sport forward and redefine the expectations for excellence. In the 1968 Olympic high jump competition, for instance, Dick Fosbury came to the games with a new paradigm and won. Up to that time, the traditional jumping style had the athlete going over the bar knee first, face down. Fosbury had perfected a new technique—labeled the "Fosbury Flop"—where he went over the bar shoulder first, face up. Fosbury and his coach had perfected a process for jumping the high bar that was fundamentally better. Anyone who wanted to stay competitive had to adopt it or lose. Today this is the only technique seen in high jump competition and it has redefined all of the standards of the sport.

Throughout, this book encourages alternative views of the product generation process, pointing to paradigms in other disciplines that might have value in the new product arena. Mapping manufacturing best practices through the information assembly line metaphor is one example of an alternate view. Exploring ways to add value to information is another. Paradigm shifts have occurred frequently in the information content that flows through the product development process. The shift to solid-state devices, the move to integrated circuits, and then the widespread use of microprocessors are some examples of these radical changes in product development content. The actual process of doing product development, however, has progressed much more slowly.

Improving development processes is an urgent task, and methods and principles are available to stimulate needed process breakthroughs. A winning framework for product development will enable engineers and their managers to enjoy the thrill of victory more often and the pleasure that comes with increasing pride in their work.

APPENDIX

A Solution to the King's Challenge

This appendix provides a complete solution to the King's Challenge introduced in Chapter 6. Basic principles that will be applied include Hartley's information flow equation given below and the fact that expected information flow is maximized when all possible outcomes from each event have the same probability.

$$I = R \, Log_2(1/p)$$

Where I is the information flow in bits per second
R is the number of events on the channel per second, and
p is the probability of the outcome from each event.

Problem Statement

To restate the problem posed in the King's Challenge:

There are ten coins on the table, identical in appearance. Nine of them are pure gold and one is false, differing from the others by weight. Also on the table is a simple balance scale. Devise a scheme whereby the false coin can be identified without fail through only three experiments with the scale.

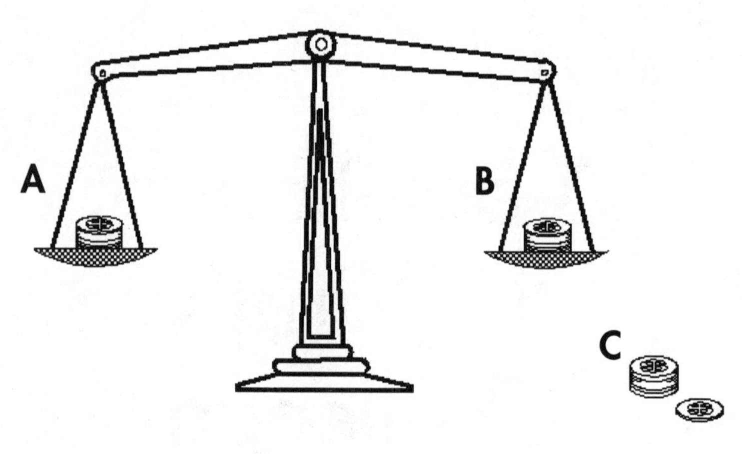

FIGURE A-1.

Experiment I.

Preliminary Analysis

To begin, the amount of information that is needed can be computed. The probability of any given coin being false is 0.1 so the amount of information needed to point with certainty to the correct coin is $\text{Log}_2(10)$ or 3.32 bits. The balance scale has three possible outcomes; tilt clockwise, tilt counterclockwise or balance. Experiments need to be designed so that the probabilities of these outcomes are as near to equal as possible to gain the most information from each experiment. If the experiments are perfectly balanced in this way, each one will yield $\text{Log}_2(3)$ or about 1.58 bits of information. Three experiments will then yield 4.75 bits which is more than enough to achieve the desired solution.

Experiment I

In Experiment I, the probabilities can be most closely equalized by dividing the coins into three groups; two groups, labeled A and B, of three coins each and one group, labeled C, of four coins. The probability that the false coin is in a particular group is 0.3, 0.3 and 0.4, respectively. The obvious first experiment is to place the two groups of three coins on the balance scale as shown in Figure A-1. The probabilities that it will tilt clockwise or counterclockwise are both

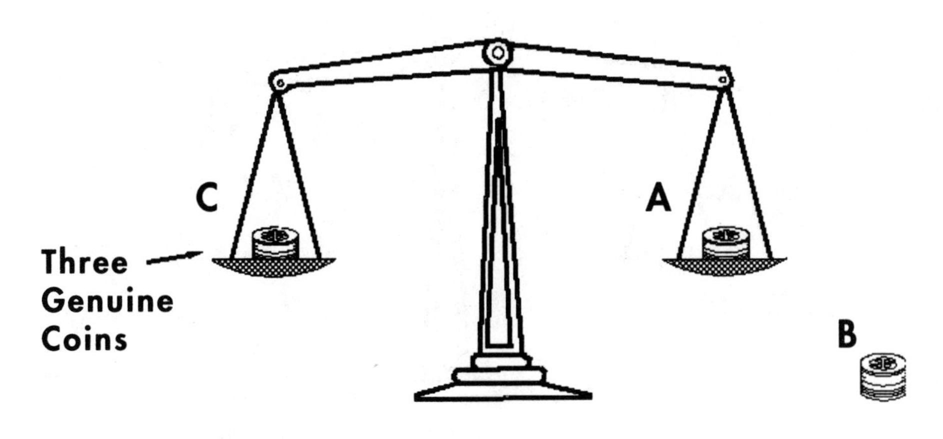

FIGURE A-2.

Experiment IIA or IIB.

0.3, and there is a 0.4 probability that the scale will balance. If the scale tilts, the experiment yields 1.74 bits of information and if it balances only 1.32 bits, but still enough to go more than one-third of the way to the goal.

The information from each possible outcome in experiment I is summarized as follows:

Ia. Tilt counterclockwise—The false coin is light and in group B or it is heavy and in group A. The coins in group C are all genuine. (p=0.3; 1.74 bits of information)

Ib. Tilt clockwise—The false coin is heavy and in group B or it is light and in group A. The coins in group C are all genuine. (p=0.3; 1.74 bits of information)

Ic. Balance—The coins groups A and B are all genuine. The false coin is in group C. (p=0.4; 1.32 bits of information)

Experiments II and III That Follow From Outcomes Ia and Ib

Experiment II requires, in general, three different strategies based upon the three possible outcomes from experiment I. As it turns out,

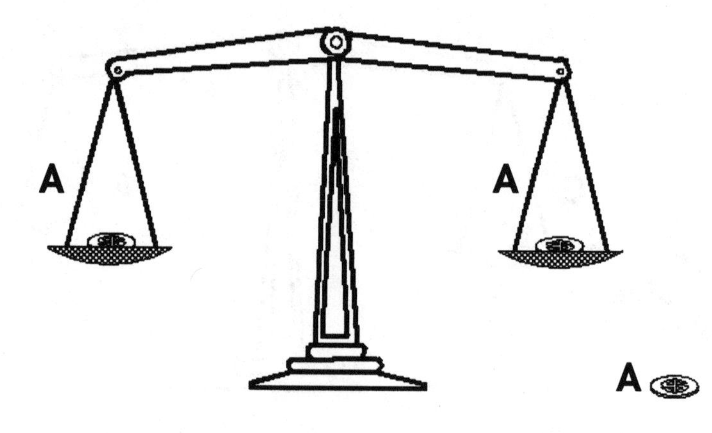

FIGURE A-3.

Isolating the False Coins from a Group of Three.

the strategies that follow from outcomes Ia and Ib are the same and will be developed first.

If the scale tilts in either direction in experiment I, then the false coin is among the six coins in groups A and B. Two coins from A on the left and two coins from B on the right would equalize the outcome probabilities at a value of one-third. This experiment requires mixing up the coins from groups A and B, however, and actually duplicates and obscures some of the information gained from experiment I. In particular, there is no way to determine whether the false coin is heavy or light from this experiment, even though the probabilities are right.

Instead experiments IIa and IIb, the experiments that follow outcomes a and b in experiment I, should both simply compare group A with three of the good coins from group C as depicted in Figure A-2. If the scale tilts at all, then the false coin is in group A and whether it is heavy or light can be determined from the outcome of experiment I. If the scale balances, the false coin is in group B, and, again, the relative weight of the coin can be determined from the outcome of experiment I. To summarize these outcomes:

IIaa & IIba. The scale tilts. The false coin is in group A and its weight relative to the other coins is determined from the outcome of experiment I. (p=0.5; 1.0 bits of Information)

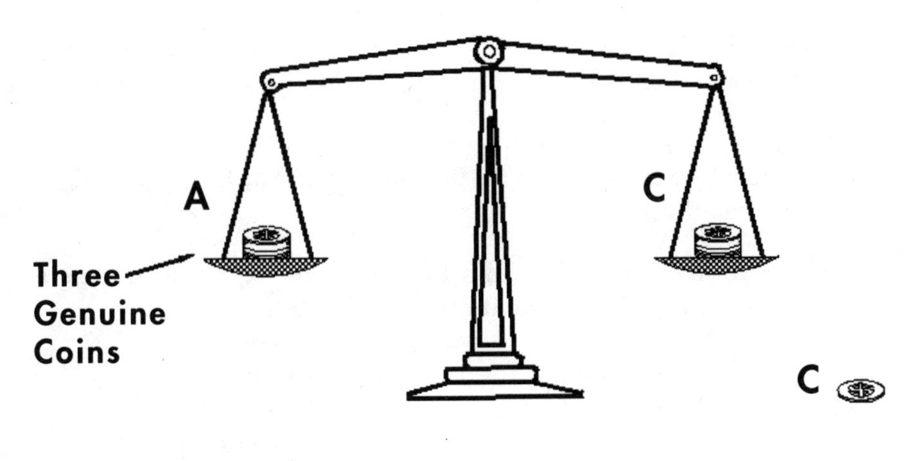

FIGURE A-4.

Experiment IIC.

IIab & IIbb. The scale balances. The false coin is in group B and its relative weight is determined from the outcome of experiment I. (p=0.5; 1.0 bits of information)

Assigning separate probabilities to clockwise and counterclockwise outcomes in this experiment is tempting. Analyzing these outcomes separately gives no new information, however, and yields an apparent additional 1.0 bits which is redundant with information gained from experiment I. When information is summed from the three experiments, redundant information should be omitted. The summary above does this and avoids potential confusion later.

The next step is to proceed to the third experiment that follows from the above outcomes and then to return to experiment I and deal with the strategy that follows from outcome Ic.

At this point, the false coin can be isolated to a group of three coins, A or B, and its relative weight is known. The third experiment is obviously to put one of these three coins on each side of the scale and leave the third on the table as depicted in Figure A-3. The outcome probabilities are equalized and the outcome identifies the false coin without fail. To summarize:

Counterclockwise: If the false coin is heavy, it is the coin on the left. If it is light then it is the coin on the right. (p=0.33; 1.58 bits of information)

Clockwise: If the false coin is light, it is the coin on the left. If it is heavy then it is the coin on the right. (p=0.33; 1.58 bits of information)

Balance: The false coin is on the table. (p=0.33; 1.58 bits of information)

The information gained from this possible sequence of experiments can be summed from the individual steps:

Experiment I	1.74 bits
Experiment II	1.00 bits
Experiment III	1.58 bits
Total	4.32 bits

This result covers exactly the 3.32 bits that were needed to identify the false coin plus one additional bit to determine whether the false coin is heavier or lighter than the genuine coins.

Experiments II and III That Follow From Outcome Ic

Returning now to outcome c of experiment I, it's clear that the false coin is one of the four coins in group C. Several versions of the second experiment are possible. Four genuine coins from groups A and B could be compared against the four in group C. This would tell us the relative weight of the false coin but would not isolate the false coin any further. Two coins from C could be weighed against two genuine coins. This would isolate the false coin further and could potentially tell us its relative weight. Another interesting variation would be to compare three of the coins in group C against the three genuine coins in group A. This comparison could identify the false coin directly or, at least, isolate it to a group of three and disclose its relative weight. Analysis of expected information return will determine which of these last two variations is best.

The expected information returned from an experiment is given by:

$$I_e = \sum_{i=1}^{i=n} p_i \, \mathrm{Log}_2(1/p_i)$$

Where I_e is the expected information from the experiment
n is the number of possible outcomes, and
p_i is the probability of the ith outcome

Applying this equation to the alternative with two coins from group C on the scale gives:

$$I_e = (0.25)(2.0) + (0.25)(2.0) + (0.5)(1.0) = 1.50 \text{ bits}$$

Applying it to the alternative with three group C coins on the scale yields:

$$I_e = (0.375)(1.42) + (0.375)(1.42) + (0.25)(2.0) = 1.56 \text{ bits}$$

Apparently, the latter of these two, depicted in Figure A-4, is the better experiment. Summarizing its possible outcomes:

IIca. Counterclockwise: The false coin is in the three from group C on the scale and is lighter than a genuine coin. (p=0.375; 1.42 bits of information)

IIcb. Clockwise: The false coin is in the three from group C on the scale and is heavier than a genuine coin. (p=0.375; 1.42 bits of information)

IIcc. Balance: The false coin is the one from group C on the table. (p=0.25; 2.0 bits of information)

The third experiment that follows from the first two results above is similar to that depicted in Figure A-3. Two of the three coins are placed on the scale and the third on the table. The three possible outcomes have equal probabilities and reveal the false coin with certainty. If the scale balances in the experiment above, the false coin is known and success has been won with only two experiments. The available third experiment can be used, however, to determine the relative weight of the false coin by simply comparing it on the scale against a genuine coin. This will yield one additional bit of information.

The information gained from outcome Ic and subsequent experiments is summarized as follows:

Outcomes

	IIca & IIcb	IIcc
Experiment I	1.32 bits	1.32 bits
Experiment II	1.42 bits	2.00 bits
Experiment III	1.58 bits	1.00 bits (optional)
Totals	4.32 bits	4.32 bits

CONCLUSION

Each path through the experiments above successfully identifies which coin is false and determines whether it is heavier or lighter than the genuine coins. Although each path involves, in general, different experiments, different probabilities and different outcomes, the calculated information yield from each path is inevitably the same, 4.32 bits as predicted in the preliminary calculations.

This problem is convenient in that the probabilities of various outcomes are easy to predict based on the number of coins involved in each experiment. Most real life problems are not so accommodating. Nonetheless, the King's Challenge serves well to illustrate the principles of information theory. It demonstrates that, at least in examples such as this, information is a tangible, calculable commodity. Even in less tractable problems, the underlying principles illustrated here can provide guiding philosophies and direction for decisions.

Index